Producer Responsibility in Practice

Also Available from Bloomsbury Academic

Handbook on Household Hazardous Waste, Second Edition, edited by Amy D. Cabaniss, PhD

Perspectives on Product Stewardship: Navigating an Extended Producer Responsibility Path to a Circular Economy, by Scott Cassel, MCP, Product Stewardship Institute (edited by Amy D. Cabaniss, PhD)

Producer Responsibility in Practice

Approaches to Advancing Product Circularity

Edited by
Amy D. Cabaniss, PhD

BLOOMSBURY ACADEMIC
NEW YORK • LONDON • OXFORD • NEW DELHI • SYDNEY

BLOOMSBURY ACADEMIC
Bloomsbury Publishing Inc, 1359 Broadway, New York, NY 10018, USA
Bloomsbury Publishing Plc, 50 Bedford Square, London, WC1B 3DP, UK
Bloomsbury Publishing Ireland, 29 Earlsfort Terrace, Dublin 2, D02 AY28, Ireland

First published in the United States of America 2025

A catalog record for this book is available from the Library of Congress.

ISBN: PB: 978-1-63671-389-2
ePDF: 979-8-7651-5478-6
eBook: 978-1-63671-390-8

Typeset by Newgen KnowledgeWorks Pvt. Ltd., Chennai, India
Printed and bound in the United States of America

For product safety related questions contact productsafety@bloomsbury.com.

To find out more about our authors and books visit www.bloomsbury.com and sign up for our newsletters.

Contents

Preface

Municipalities have been left "holding the bag" for consumer products discarded at their useful or desired end-of-life (i.e., post-consumer). They have had to contend with the administration and cost of municipal solid waste collection and transport, management, and disposal, stressing the budgets of municipalities to accommodate the ever-increasing abundance of post-consumer product and packaging waste. The overconsumption of goods and a throw-away mentality are reflected in high rates of municipal solid waste generation and disposal. With finite natural resources and increasing population, this linear approach of "take-make-waste" is deleterious to the environment and unsustainable.

Circular processes are regarded as a viable alternative to a linear economy. In a circular economy, "we stop waste being produced in the first place."[1] Rather than burial in landfills or incineration at product end-of-life, materials are collected, processed, and re-entered into the system through reuse, recycling, refurbishment, and remanufacturing. Manufacturers and brand owners ("producers") play pivotal roles in determining the design and composition of the products they produce and can to an extent predetermine the destination of products at end-of-life (EOL). They have a responsibility for the products and packaging; other stakeholders (e.g., industry, government regulators, retailers, and recyclers) share responsibility.

To decrease the amount of post-consumer waste and the burden of waste management borne by municipalities, responsibility needs to shift to manufacturers and brand owners, with oversight and regulation by government. This shift is underway in the United States through the adoption of Extended Producer Responsibility (EPR). EPR is "a law that extends a producer's financial and managerial responsibility for its products beyond

[1] Ellen MacArthur Foundation, "What Is a Circular Economy?" 2023, https://www.ellenmacarthurfoundation.org/topics/circular-economy-introduction/overview.

the manufacturing stage—both upstream to product design and downstream to post-consumer reuse, recycling, or safe disposal." This includes "the steps needed for a material to go from consumer waste to a new product or package."[2]

In many EPR models worldwide, and in most in the United States, a nonprofit Producer Responsibility Organization (PRO) is set up, funded, and managed by an industry trade association and the manufacturers or brand owners. Once a PRO plan is approved by the governing state agency (which maintains oversight), the PRO independently, or in coordination with other PROs, provides administration to fulfill obligations and achieve designated outcomes in a statewide EPR program for specific product and packaging waste streams (i.e., covered products). Obligations include such things as collection (e.g., take-back programs) and responsible processing of post-consumer recyclable materials, meeting recycling rate targets and reimbursing costs to local governments and processors. Municipalities can continue waste management program operation (e.g., collection of recyclables) and other roles such as providing community outreach and education. Mandatory EPR schemes help to ensure sufficient and sustained funding, as well as other benefits, "such as enhancing the efficiency and transparency of the system and incentivizing upstream solutions."[3]

EPR policy can, for example, incentivize producer attention to "eco-design" practices that minimize adverse environmental impacts. This includes design for ease of deconstruction for recycling and remanufacture, incorporation of post-consumer material into a product, and the use of less material, as well as less toxicity. "If producers were made responsible for end-of-life management (i.e., reuse, recycling, energy recovery, treatment, and/or final disposal) of products, they would find it in their self-interest to anticipate end-of-life costs and obligations and design their products to minimize those costs."[4] Internalization of the costs could, in the end, translate to cost-savings

[2] S. Cassel, *Perspectives on Product Stewardship: Navigating an Extended Producer Responsibility Path to a Circular Economy* (Boston: Product Stewardship Institute, 2023), p. 48.

[3] Ellen MacArthur Foundation, Extended Producer Responsibility: A Necessary Part of the Solution to Packaging Waste and Pollution. 2021. https://www.ellenmacarthurfoundation.org/extended-producer-responsibility/epr-statement (accessed 2023).

[4] R. Lifset and T. Lindhqvist, "Producer Responsibility at a Turning Point?" *Journal of Industrial Ecology* 12 (2) (2008): p. 144.

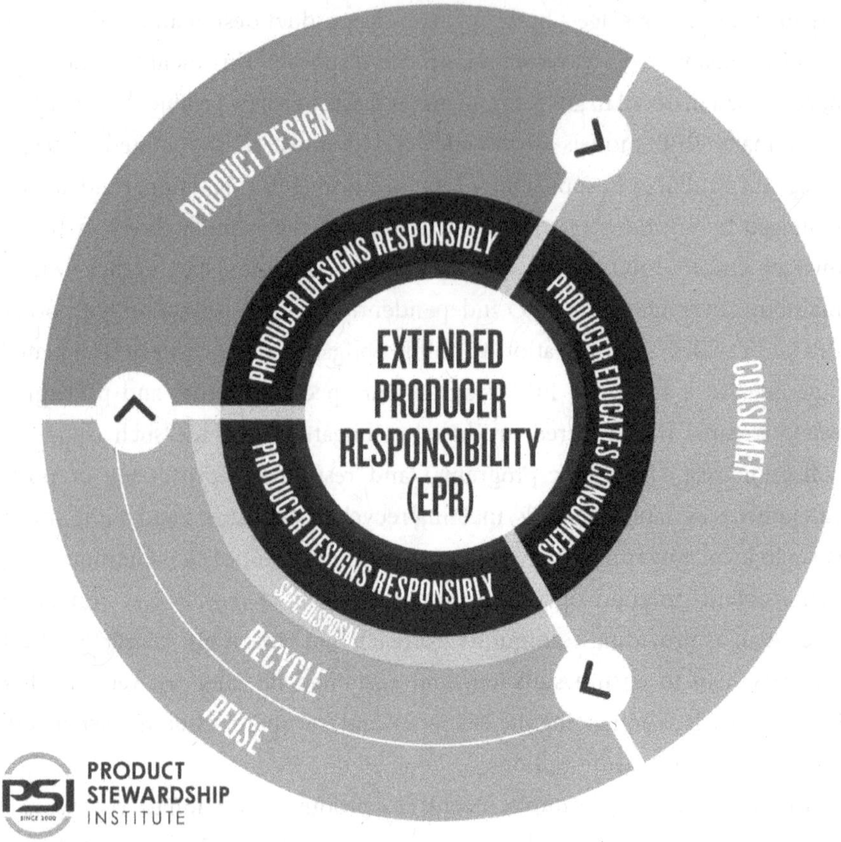

Figure 0.1 Extended Producer Responsibility: Product design, consumer engagement and education, and product waste management.

Source: Product Stewardship Institute, 2023, https://productstewardship.us/what-is-epr, reprinted with permission.

or price reduction for consumers.[5] This aspect of EPR—incentivization for design for environment (DfE) or design for recycling (DfR)—appears to need strengthening both in assessment and actualization, however.[6] Through EPR legislation, producers are expected to assume responsibility throughout their products' lifecycles, including product design and financial responsibility for their products at EOL (see Figure 0.1).

5 Lifset and Lindhqvist, p. 144.
6 OECD, *Extended Producer Responsibility: Updated Guidance for Efficient Waste Management* (Paris: OECD Publishing, 2016). https://www.oecd.org/environment/extended-producer-responsibility.htm# (accessed December 16, 2023).

EPR in the United States follows on the heels of EPR progress in Europe and Canada. More and more US states are adopting EPR legislation and expanding the types of covered materials. This is a fast-changing landscape. This super-brief preface does little to shed light other than to offer a glimpse into EPR and begin to set the stage for the chapters that follow in *Producer Responsibility in Practice*. In addition to the wealth of knowledge and experience shared by chapter authors in this book, you would be well-served to regularly seek information on EPR and the circular economy as it continues to evolve. As editor of *Perspectives on Product Stewardship* by Scott Cassel, Product Stewardship Institute (2023), I can attest to the value of this book in particular, as it provides a strong foundation on the evolution of EPR and its application in the United States with leadership and facilitation by PSI and others. My gratitude to Scott for his extraordinary, sustained efforts.

Producer Responsibility in Practice

Producer Responsibility in Practice was developed with the intention of sharing examples and insights on EPR offered by those actively engaged in it. Chapter authors are experts who have "been there, done that," and continue to work on promoting the circular economy and EPR legislation, developing public–private partnerships for EPR-covered products, designing and implementing community-based programs to motivate consumer participation in waste management and EPR initiatives, and more. My sincerest thanks to the authors for their valuable contributions to this book. In the pages that follow, these leaders share their knowledge, perspectives, and experiences. They offer examples and suggestions that can help readers rise "higher on the learning curve" by sharing challenges, failures, and successes that can serve to inform those striving to implement or augment EPR programs in their states. Questions such as, how have others done it? What are some legislative approaches to gaining traction? How can communities and Producer Responsibility Organizations or Stewardship Organizations engage consumers in program participation to capture products and packaging at end-of-life? These are some of the many considerations and perspectives addressed in this book.

Product Responsibility in Practice was conceived as an informational and educational guide for decision-makers who seek to lay the groundwork, or build the infrastructure, for EPR on products such as mattresses, paint, pharmaceuticals, and household hazardous waste in their own states. Chapter authors share their perspectives that serve both as cautionary tales and stories of progress that have opened doors to opportunities for furthering EPR in the United States. While many of these experiences and outcomes are unique to the authors, geographic locations, and specific covered products, they offer valuable understanding and "tools" for solid waste managers, legislators, producers, program administrators, and others who strive to reduce environmental impacts throughout product lifecycles and make a difference in their communities, now and in the future.

In Chapter 1, Kristin Aldred broadly considers consumer product design, use, and disposal, advocating for a systems perspective and shared responsibility for products throughout their lifecycles. Dr. Aldred shares insights obtained through research literature on behavioral constructs and interventions to encourage individual behavior and societal change.

What started as a low-tech latex paint collection and recycling program administered by Metro in Portland, Oregon, a regional government, blossomed into a successful paint industry-operated system for the collection and management of leftover paint. In Chapter 2, Jim Quinn takes us from the early days of their operations through the development of the public sector–private industry partnership and Extended Producer Responsibility (EPR) for paint, and eventually the 2009 Oregon legislative session when the first paint stewardship bill was passed in the United States.

Chapter 3 tackles EPR head-on. Authors Heidi Sanborn and Tim Goncharoff share their experiences and insights on working with stakeholders in the introduction, passage, and implementation of EPR legislation for various product types in specific circumstances. In addition to presenting several challenges, the authors have constructed this chapter so that it offers many valuable recommendations for approaching EPR bills.

Chapter 4 looks at mattress management. Part One of this chapter is authored by several members of the Mattress Recycling Council, a stewardship organization established by the mattress industry to fulfill its EPR obligations.

They detail the composition of mattresses for material recycling and outline how mattress recovery and recycling work. They also describe industry workgroups whose efforts extend beyond consideration of materials recovery to other aspects of the product life cycle.

In Part Two of Chapter 4, state agency official Tom Metzner shares several driving forces behind Connecticut's success as the first state in the United States to pass EPR for mattresses. Mr. Metzner describes ways in which the program has grown with municipality engagement, indicating the Connecticut mattress EPR program has been "a resounding success."

Consumers share the responsibility for waste minimization and pollution prevention through individual efforts such as participation in product take-back programs. In Chapter 5, Jennifer Tabanico presents research on motivating consumer behavior through community-based social marketing campaigns. She describes the steps involved in CBSM and its demonstrated value in fostering effective behavior change programs. This chapter serves as a valuable guide with instructions, prompts, examples, and resources for those within municipalities, stewardship organizations, and other entities who are responsible for overseeing or developing and implementing impactful public programs.

Concluding remarks are offered by David Nightingale wherein he considers several of the main points addressed in the book and provides his perspective on the past, present, and future of EPR in the United States.

With recognition that resources are finite and often retain value at product end-of-life, and that pollution can be prevented throughout the product chain, it is ever more important to implement product stewardship practices and Extended Producer Responsibility policies for sustainability. We have seen that EPR continues to grow in the United States, as it has elsewhere. Our hope is that, from demonstration on these pages of EPR in practice from various viewpoints, you can become better informed about potential challenges and approaches. In addition to being inspired in your pursuits, you will be better equipped to support and grow the efforts of others working in this realm. Regularly tap into the vast array of resources available to further guide and support you in your efforts. Thank you. Cheers to your past and future successes!—Amy Cabaniss

Disclaimer

The information provided in this book is for general informational and educational purposes only. Readers are encouraged to consult with qualified professionals for advice tailored to their specific circumstances and needs before making any decisions based on the information provided herein.

This book is a compilation that reflects the diverse perspectives and opinions of its respective authors, not necessarily those of the other contributors, editor, or the publisher. The publisher, editor, and authors make no warranties or guarantees, expressed or implied, about its accuracy, completeness and sufficiency, or reliability. The publisher, editor, and authors disclaim any and all liability for any errors or omissions, and for any damages, including but not limited to direct, incidental, consequential, or special damages arising out of or in connection with the use of this information.

1

Systems-Based Design with Resource Recovery in Mind

Kristin Aldred

Designing Our Way Out of Complex Problems: Use of Multiple Approaches to Affect Change

How many people does it take to change a lightbulb? A lot, it turns out.

At the beginning of my career, while conducting large-scale commercial energy audits, I counted tens of thousands of light bulbs in offices, schools, manufacturing plants, and even an aquarium. Once the audits were complete, we presented our clients with money-saving solutions to reduce energy use. Lighting was easy to address with *technological innovation*, which simply meant replacing incandescent bulbs with fluorescents and later LEDs. Lighting projects had a quick payback. For a modest investment, clients were virtually guaranteed to break even within three or so years, and then the savings would start adding up. Despite the potential financial and energy savings, organizational leaders were surprisingly resistant to implementing our recommended changes.

Direct savings from installing more energy-efficient bulbs was just one part of the equation. Utilities needed to reduce electricity demand to delay massive capital outlays to expand infrastructure. We explained to our clients that there were also *financial incentives*. Rebates from utilities offset an organization's investment and further reduced payback times for lighting projects. From a financial perspective, the logic in our spreadsheets was undeniable.

Still, savings from new technology and direct incentives were not enough to overcome many objections from key stakeholders. Installing new lighting came with opportunity costs, such as redirection of staff from other projects, and short-term disruptions. These concerns were harder to quantify or address by sharing payback calculations. As consultants, we needed to engage with different stakeholders to *educate* and to find ways to *motivate* them based on their reasons for resisting change in the face of apparent benefits. For example, for staff in health and safety or human resources, we might highlight the relationships between new lighting and safety or productivity. For staff in facility management, we might focus on lower maintenance requirements with new lighting technologies, which could free up staff time for other, perhaps more critical, tasks.

After a client decided to move forward with a project, we turned our attention to behavior. We understood, and our clients confirmed, that people would leave lights on when not needed, wasting energy and money. We worked with them to build *awareness*, provide information, and tie individuals' efforts into a broader "green" campaign to persuade building occupants to turn out the lights. Even with good intentions, people need reminders to turn off the lights and so we used point-of-behavior prompts such as "Please turn out the lights" stickers that were placed by the light switches. Prompts are useful, for a time.

We offered a technological solution—*automation*—in the form of motion detectors. With automation you don't need to rely on people remembering to do something or developing new habits. Automation takes behavior out of the equation, although it does have its limits. Anyone who has sat in a dark space, wildly waving their arms to trigger a light motion sensor, knows that the technology must meet people's needs, when and where they are. Anyone who has seen a cold, wet paper towel draped over a locked thermostat also knows that people will find a way to circumvent technology if it's not meeting their needs. Today, more sophisticated smart buildings can detect how many people are in a room, adjust or shut off lights, and even accordingly fine-tune ventilation. Heating, ventilating, and air conditioning (HVAC) can be remotely programmed room by room to fit unique use patterns.

Beyond the tools described, we can also use *policy* to affect change. People who resist change are not necessarily against it. Education, incentives, and other voluntary measures can fail simply because there are other, and sometimes more pressing, responsibilities and priorities. Policy can be an important tool to bring about public benefits such as protecting health and environmental quality. In the case of lighting, a policy was eventually implemented to phase out incandescent bulbs.

From this example, it's obvious that changing light bulbs to conserve energy takes more than people installing new equipment. It can take many people including engineers, analysts, financial experts, communication strategists, behavioral psychologists, and policymakers working together. Understanding this and seeing the complexity of solving a seemingly straightforward engineering and financial problem deepened my interest in psychology and behavior change as part of a larger system. I began to see how placing the burden on "the individual" to know and remember what to do in hundreds of situations each day was not sufficient or effective. Through new products and environments, we can better support people in their efforts.

Learning about a safety procedure a bit later in my career brought the idea of using design to change behavior into focus. Lockout-tagout is a procedure for temporarily disabling equipment to protect workers from hazardous electrical, hydraulic, mechanical, and other energy sources that can cause serious and even fatal injuries. When machinery is shut down for maintenance or repairs, a lockout device or "hasp" prevents someone from turning on the equipment and unknowingly endangering workers who are conducting maintenance. A hasp can have slots for several padlocks, so each worker must remove their lock before the equipment can be started up. A "tagout" is a small sign attached to the hasp that alerts people of the danger of restarting the equipment (see Figure 1.1).

The lockout device was designed specifically to force a behavior when life and limb are at risk. Awareness, education, persuasion, building a culture of safety, and other methods play vital roles, but a lockout device is a built-in fail-safe. Its use is so important that it is codified in regulation [OSHA standard for The Control of Hazardous Energy (Lockout/ Tagout) (29 CFR 1910.147)]. Lockout-tagout is an example of extensive steps we take to direct behavior when the stakes are high.

There are far less critical behaviors that have been changed by the impact of design and other technological advances. Many of our behaviors that have

Figure 1.1 A hasp, lock, and tagout to prevent equipment from being turned on and endangering workers.

Source: Creative Commons license.

been enhanced or completely changed in the past few decades, such as the ways that we listen to music, take photos, and retrieve information. Companies like Apple not only created new technologies but also facilitated behavior change by introducing products that met people's needs in ways that were easier, more convenient, and often more delightful to use than existing products. As behaviors changed, many products became obsolete, and with little fanfare as they faded away. Fundamental behavior changes were not driven by educating or persuading people to change but by new products and environments designed to support change.

Today we face large-scale environmental degradation that threatens the natural systems on which we, and many other species, depend. Yet our approach to addressing this destruction has tended to fall into one of two categories: *incremental changes*—many of which have lingered on the agenda for decades and failed to keep pace with the rate of change in the problems we face—or the *status quo*, maintaining and continuing along the same trajectory while either denying a problem exists or believing that nothing can be done about it.

Given the complexity and scale of our environmental problems, the only viable path over the long term is a third option: to make *systemic change*. Not piecemeal progress or continuance along a familiar, worn path. Intentionally or not, we designed ourselves into the current situation. We need to *intentionally design our way out* of it.

What does this mean and why is it being addressed in a book on shared resource management and Extended Producer Responsibility? The products discussed in this book were designed by someone. While definitions vary, one way to think about design is that it's an intentional process (Blackler et al., 2021). Good design is about problem-solving and envisioning how things ought to be. Ideally, design is holistic, meaning the process takes into account the resources used, where they came from and where they go. This includes the people who will use the product and how they'll use it; the environment, broadly speaking, where the product will be used; and what will happen to the product when it is no longer useful. That last part—what happens to products when they're used up or unwanted and discarded—is still often overlooked in the design process. Yet the way a product is designed has tremendous impact on how it is managed at product end-of-life, including whether and how easily the resources embedded in it can be recovered.

My intent with this chapter is to bring the people working on different sides of the product life cycle closer together. Some basic questions about the design of consumer products and their resource value need to be discussed. We'll consider the levers that can be pulled to bring about change in a system beginning with the early stages in product development. We'll examine why it's time to stop heavily blaming consumers for systemic failures. Although a major focus of this chapter is on wise resource use and recovery from consumer products, there are other, inseparable issues around waste and the efficient use of resources that will be considered. My hope is that this chapter will encourage you and others, such as those who work in policy, product design, communications, education, behavior change, manufacturing, and other realms, to broaden your perspective and give attention to whole systems thinking, moving away from a narrower focus on individual materials, products, or stages of a product's life cycle. The best-case scenario encourages active engagement with one another to solve some of our problems, be they small or big. One way to do this is with an eye toward product stewardship that prioritizes collective resource stewardship.

Product Stewardship Is Resource Stewardship

Everything operates in a larger system. We can think of ourselves and our unique characteristics, beliefs, capabilities, and even motivations, as being within the system where we interact with products and our physical environment. We are shaped by, and influence, our families, workplaces, and communities. We help to determine, and are subject to, laws and regulations. Enveloping it all is our culture plus associated norms and customs. The system isn't stationary in time—what happened before impacts us now and our decisions will impact the system in the near and distant future.

The light bulb example shared earlier demonstrates that successful change requires an expansive view, one that reaches beyond a narrow part of a system. Using energy audits as a gateway, our consulting firm worked to foster organizational change on a larger scale. Together we sought to influence individuals' decisions, alter schedules to reduce resource use, develop policies to implement, support and institutionalize change, and encourage leaders to change the culture of the organization to one that acknowledged and prioritized responsible resource use. It wasn't just about energy, rather it was inclusive of considerations regarding water, waste, manufacturing processes, and the products themselves.

Products today are largely designed to be consumed in a linear "take, make, waste" process, where we take resources, design, and make a product, and the product is eventually disposed, usually as waste. We know that resources aren't fully *consumed*. Resources are stored for some period in the form of products and these "man-made" objects that people interact with every day—from buildings to bicycles, bowls, and bags—are basically our natural environment combined in new ways, designed to meet particular needs and wants.

Rarely do all the resources used to make a product actually go into the product itself. The production process, for example, typically results in waste material. It also generates waste byproducts that can pollute our air, water, and land, sometimes imperceptibly and other times to devastating effect. More waste and pollution are generated when we use products. When products reach the end of their useful life, unless an effective process is in place to recover and reuse or reprocess materials, the products themselves become waste.

Producers benefit by converting resources to products, and consumers benefit from the actual or perceived usefulness of those products. But then typically *everyone* shares in the cost of managing, and bearing the impact of, "externalities." Externalities are the responsibilities and impacts associated with a product that are redistributed from producers and consumers to everyone. In a "take, make, waste" economy, waste is "external" to the producer and the consumer. Externalities are somewhat easy to ignore in a linear process. Operating along a straight line, like looking through a spyglass, you don't see what's happening at the margins and beyond—in either direction—or pay attention to how things might be connected today, or over time.

The reality is that taking, making, and disposing of materials and products at end-of-life don't happen in a vacuum. It happens *somewhere* inside a system. Resources we take come from the same system and waste is generated through activity inside that collective and codependent system, where it remains. Products are undeniably part of the same broad system on which life depends.

From our perspective, the purpose of the system in which we live is to meet our needs, wants, and desires. Hopefully we recognize the interdependence among people and other species and therefore recognize that the purpose of the system is also to meet their needs. Hopefully we also believe that the system should keep serving this purpose in perpetuity, with a recognition that resources are not inexhaustible. Along with our understanding that products are part of this system, it's easy to see why keeping materials circulating in the system in useful ways and protecting the remaining resources from degradation are very important. Every ounce of waste or pollution had at some stage the potential to be useful or provide more benefits.

It's clear that resource recovery needs to be "built in" during the very early stages of product design. Recognizing the need for resource recovery without designing recovery into the production of goods would be continuing to follow a linear process while acknowledging that there's a larger system, yet only taking responsibility for the part you can see. It's no longer good enough to make decisions in one part of the system and leave it up to someone else in another part of the system to deal with the consequences of your decision—especially those who had no input in your decisions or actions.

While much of the emphasis in this book is on product management at end-of-life, it keenly focuses on responsibility, recognizing that product stewardship encompasses much more than what happens at that stage. It's not "just" resource reuse and recycling. Product stewardship also comprises an ethical commitment from companies to take responsibility for the management of their products throughout all stages of manufacturing and distribution (Fiksel, 2012). Ultimately, product stewardship is resource stewardship, and efficient resource use is critical to sustaining a healthy, thriving system.

Product stewardship starts with thinking about how products are designed, manufactured, used, and managed inside the large, complex system on which we all depend. While we are somewhat better today at systems-thinking, we have much further to go in terms of design and of fairly distributing and accepting responsibility for resource acquisition, use, and recovery.

Design for the Environment and Sustainable Behavior

Design for the environment (DfE) focuses on product innovation. DfE innovation enables "continued growth and prosperity" while at the same time reducing a company's environmental footprint (Fiksel, 2012). A fine example is Caterpillar, Inc., which manufactures construction and mining equipment, engines, turbines, and locomotives. Caterpillar is a pioneer in remanufacturing, having first established a facility used exclusively to remanufacture truck engines fifty years ago. The company has since built a worldwide product take-back program. Its remanufactured products produce 65 to 87 percent less greenhouse gas process emissions, use 80 to 90 percent fewer materials, and use 65 to 87 percent less energy than new products. In 2022, the company took back 150 million pounds of materials (www.caterpillar.com/en/company/sustainability/sustainability-report.html).

To create and sustain an effective take-back and remanufacturing program at such a large scale, Caterpillar has to contemplate the entire life cycle of a product from the very beginning of development and design. The type of materials used, the ease with which a product can be disassembled and repaired, and the ability to reclaim materials all have to be considered. With their careful strategy, Caterpillar is able to keep materials circulating through multiple uses. As they state simply, "Caterpillar products are built to be rebuilt" (www.caterpillar.com/en/company/sustainability/remanufacturing.html).

In addition to DfE, design for sustainable behavior offers an approach for companies interested in curbing resource consumption, particularly during a consumer product's use phase. Design for sustainable behavior is a user-centered research framework that employs knowledge of how people use products to identify design solutions for shaping behaviors in environmentally sustainable ways (Boks, 2012; Selvefors, Pedersen, and Rahe, 2011). Unlike other sustainable design approaches that focus on the efficiency, reduced toxicity, material content, reusability, or recyclability of products, design for sustainable behavior focuses on consumption during the use phase of products ("'Designing-in' sustainable behavior," 2008; Lilley, 2009; Lockton, Harrison, and Stanton, 2008; Tang and Bhamra, 2009). Designers aim to change behavior through two general means: by providing feedback that makes consumption more tangible or by scripting or steering behavior (through introducing obstacles, providing guidance, or making a preferred behavior easier with affordances). Changing behavior requires an in-depth understanding of how people interact with their environment, particularly with products.

Pulling Levers in the System for Resource Stewardship

Just as waste comes in a lot of forms, there are a lot of ways to reduce it. In a circular, rather than linear, product life cycle, we "see more": more steps, more impacts, more opportunities (see Figure 1.2). From resource extraction through manufacture, distribution, use, and recovery, there are opportunities to design-in proper waste minimization and resource recovery. The key is that product stewardship—or minimizing the impacts and maximizing the benefits of a product and its packaging—is in every step of the cycle. Product stewardship isn't an activity that occurs only at the product's end-of-life. The decisions made regarding how the product is designed, formulated,

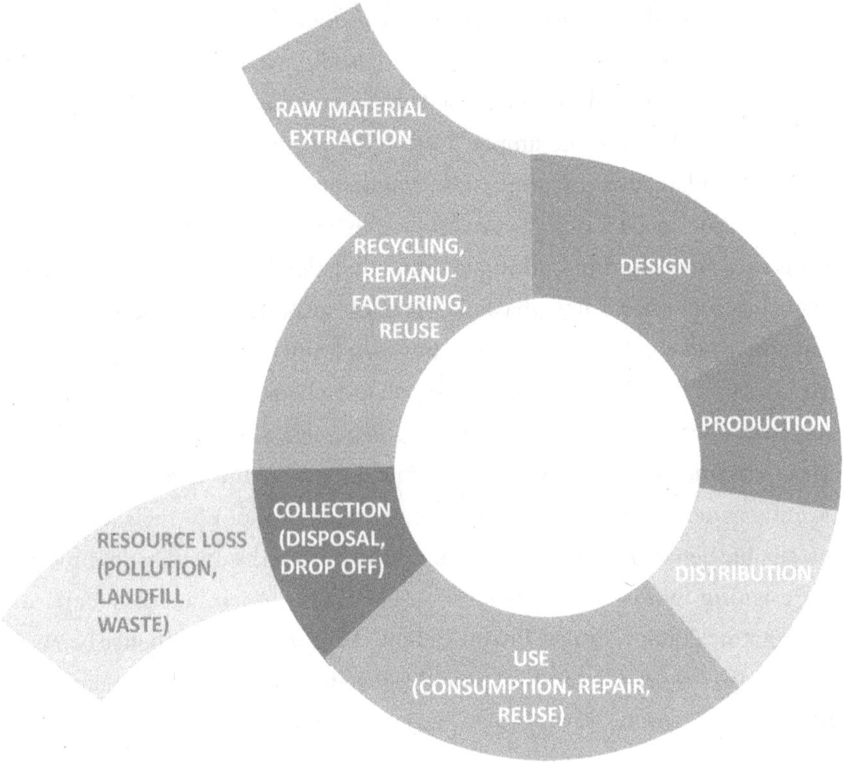

Figure 1.2 Simple life cycle diagram.

Source: © Kristin Aldred 2018.

manufactured, and transported all impact the need and ability to recover materials later. Product designers and the companies that employ them are not the only ones with product stewardship responsibilities. Government has responsibility for setting guidelines and boundaries, ensuring a level playing field for producers to prevent one player's actions from having either an unfair advantage or placing an unfair burden on another. Each stakeholder bears a responsibility for product choices including how we use and dispose of those products.

Within the life cycle of a product, there are three major phases when we can intervene to reduce a product's impacts: design, production (which includes raw material extraction, manufacture, distribution), use, and end-of-life disposal or recovery. Design innovation is an important part of Extended

Producer Responsibility. Design innovation is especially critical because it impacts every other phase of a product's life cycle, making it an essential component of effective product stewardship.

In this book, you'll read about product stewardship and EPR for paint. A paint producer's decisions matter regarding product formulation, both in terms of the up- and downstream impacts of resources used in the paint. So, too, does the quality of the paint: it affects how much is needed for the job at hand and how long it will last. Paint packaging (e.g., cans) has implications for shipping-related resource use and the recoverability of resources related to that packaging, as well as the function it plays in preserving leftover paint that can be recycled. Paint recovery and recycling require several supporting factors, starting with attention to the original product formulation, consumer education about proper storage, the infrastructure needed for consumers to conveniently return unused (left-over) paint, policies to support the recycling industry, and, in some cases, policies compelling collection and recycling. Stakeholders share responsibility in paint stewardship, but in many stages of the life cycle, the producer has the most influence and therefore the greatest responsibility. EPR codifies this responsibility through policy by mandating that a manufacturer's responsibility for a product extends to post-consumer management of that product and its packaging. EPR shifts financial and management responsibility upstream to the manufacturer and away from the public sector while also providing DfE incentives to manufacturers (Product Stewardship Institute, https://productstewardship.us/what-is-epr).

One essential stakeholder group is overlooked until the end of the life cycle: people in the solid waste management industry who are working to responsibly solve problems they didn't create. The waste industry is usually disconnected from production. With product design and manufacture, planning for a product's end-of-life management appears to be typically treated as a separate issue. Despite the need for the expertise of solid waste management experts, they don't usually have a seat at the table early in a product's life cycle. Waste experts are typically handed things late in the game, when there are few good product stewardship options, and sometimes, only complications. Take for example the Hood Lightblock Bottle® that was introduced in 1997 as a new milk jug designed to reduce light exposure to minimize milk quality degradation. This opaque, white-pigmented plastic packaging was used instead of translucent high-density polyethylene (HDPE, #2) bottles which are frequently collected in community recycling programs. Its introduction was of concern for reprocessing with other plastics due to lower market value and higher cost to recycle (Smith, 1998). The issue was large enough for a recycler to contact dairies throughout the region that might be considering a similar change in packaging. One dairy producer, Garelick Farms, responded by consulting with recyclers and following recommended guidelines when designing new packaging. This situation shows how having solid waste management experts "at the table" can be beneficial early in the process. While there are more efforts today to integrate packaging design and recyclability (e.g., the Sustainable Packaging Coalition), this is not the case with product design and resource recovery more generally. Because resource recovery is important for your product/brand, involving solid waste experts in the design phase is important, just as you might consult a materials scientist to choose less-toxic ingredients in product manufacture.

The Special Case of "Consumers"

Traditional Behavior Change Efforts and Behavioral Economics

The history of conservation behavior such as recycling, composting, reuse, and energy conservation reflects a tendency to put much of the responsibility for waste management on consumers. It's up to us to deal with products, packaging, and organic materials. When things aren't working (e.g., low recycling rate), blame is placed on consumers. It is implied that people don't care, there's a lack knowledge and understanding, or maybe people aren't motivated and can't be inconvenienced. While part of the problem should of course be attributed to individuals, when you look at the life cycle of a product and consider the larger system in which everything operates, you see how unproductive it is to blame individuals for our systemic failures related to waste and resource recovery.

Enormous investment is made by companies in the logistics for distributing products. Yet for many products or brands, a lack of investment in collection becomes readily apparent at product end-of-life. More attention to this aspect of product stewardship and EPR is needed early on. To transition from a linear to a circular economy, consumer behavior remains of utmost importance in the collection of leftover and unwanted products. In asking people to return products for reuse, recycling, and resource recovery, we need to build in the capacity and support for the behavior at the outset or we won't be successful asking people to act. With the necessary infrastructure in place for collection of discards (e.g., through municipalities, retailers), people must be encouraged to participate in collection programs because, ultimately, the recapture of materials from products at end-of-life rests disproportionately on the shoulders of consumers.

Research literature can be tapped into to inform interventions to change individual behavior; it focuses on various constructs such as awareness and knowledge, values, beliefs, and attitudes toward environmental problems. Approaches that have been effective in increasing pro-environmental behavior include employing persuasion techniques, providing feedback, and inducing norms in a target audience. Often, messages are framed in terms of altruism, morality, or sacrifice (De Young, 2000; Gardner and Stern, 2002; Kaplan, 2000). The most effective interventions typically use some combination of strategies (Gardner and Stern, 2002; Schultz, 2014; Steg and Vlek, 2009; Stern, 2000). Given the scale and continued proliferation of environmental problems and consequent impacts on human health, plus known persistent gaps between attitudes and behaviors and between intentions and behaviors (Kollmuss and Agyeman, 2002; Kurz, 2002; Winter and Koger, 2004), we need new ways of thinking about individual behavior and consumers' roles and responsibilities relative to product stewardship. (See also Chapter 5 by Tabanico in this book.)

Research literature on human behavior shows persistent gaps between attitudes and behaviors and between intentions and behaviors (Kollmuss and Agyeman, 2002; Kurz, 2002 Winter and Koger, 2004). Where people's stated or inferred best interests and their actual choices (behaviors) don't line up (Akerlof and Kennedy, 2013; Thaler and Sunstein, 2003), behavioral economics interventions often employ "choice architecture" (e.g., nudging) to alter the decision context while preserving a person's ability to choose (Akerlof and Kennedy, 2013; Sunstein and Reisch, 2014; Thaler and Sunstein, 2003). The desired behavior for example, can be the "default option" wherein intentional design influences the person's choice. The most common type of intervention is the use of a default option (Croson and Treich, 2014; Reisch and Sunstein, 2014; Sunstein and Reisch, 2014). Because of the status quo bias, people tend to stick with the default option. People are also sensitive to losses and may perceive movement away from the default as a loss. For example, in two real-world settings, utility customers who were offered electricity from "green" (i.e., renewable energy) sources as the default option, kept the source selection (Pichert and Katsikopoulos, 2008). In lab experiments involving electricity choice, participants also were more likely to stick with the default, even if it was more expensive. Small changes using default options could potentially have a greater impact on environmental quality than economic incentives, education, or moral appeals (Reisch and Sunstein, 2014), although it usually takes a combination of interventions to achieve success in affecting behavior change.

Behavioral economics essentially considers our natural ways of thinking—including systematic errors and our use of mental shortcuts—to improve individual and sometimes collective well-being (Sunstein and Reisch, 2014). In addition to making the desired behavior the default option, infrastructure and product placement play key roles in behavioral choice. Placement of a recycling bin next to a trash bin influences the decision to recycle; bins with shaped openings (e.g., a circle for bottles or slot for paper) encourages proper sorting; and a cigarette filter receptacle attached to a trash bin encourages sorting, providing a convenient bin (and prompt) for individuals to dispose of cigarette butts separately from regular trash.

Durable bag use in Canada became more widespread when a major grocery store chain took a new approach to the bags and their placement (Nisbet and Gick, 2008). When the store began selling bags with a "stylish" design with easy access at the point of checkout, consumers were more inclined to choose them. Within the product design field, there is interest in proactively using design to facilitate pro-environmental behavior (Boks and Daae, 2017; De Medeiros et al., 2018).

Environmental Psychology

Returning products at end-of-life is a complex behavior that warrants further study. Environmental psychologists study behavior in everyday setting and

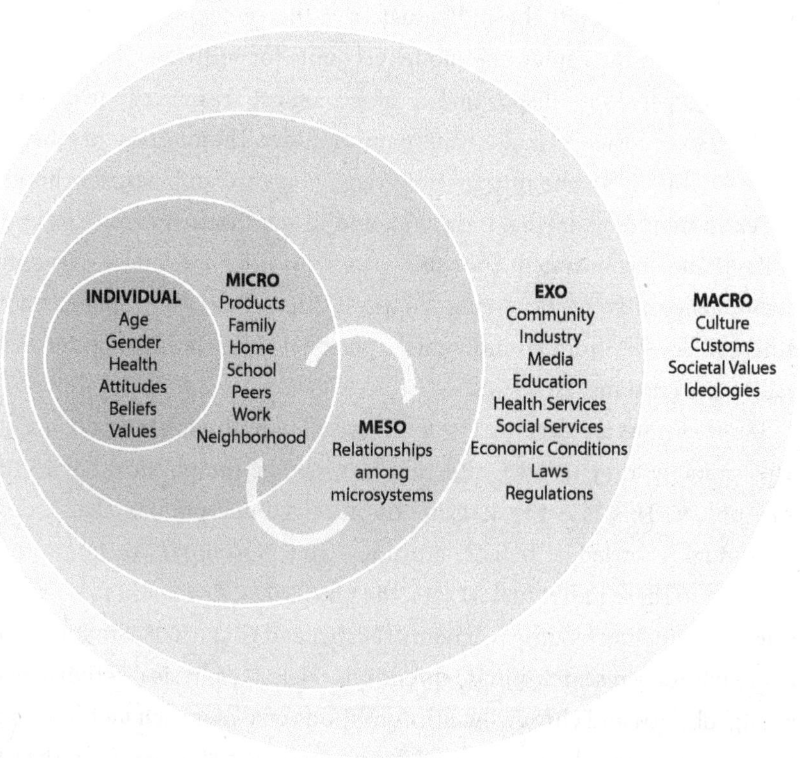

Figure 1.3 The bioecological model.

Source: Adapted from Bronfenbrenner.

are oriented toward problem-solving and using systems models to better understand people. The field of environmental psychology is well-suited for helping to address environmental sustainability and circularity (Gifford, 2007).

The bioecological model (see Figure 1.3), which was adapted from the field of human development (Bronfenbrenner and Morris, 1998), provides a useful framework for understanding complex systems and multiple factors that interact and that influence human behavior. This model has been applied effectively to public health issues, for example. Public health interventions today take a systems approach to impact behavior on a community level (Booth et al., 2001; French, Story, and Jeffery, 2001; McLeroy et al., 1988; Razani and Tester, 2010; Sallis 1998; Stokols, 1992a; Story et al., 2008; Wells et al., 2007).

The model is presented here because it illustrates that which was described earlier in this chapter: the individual is at the center of a series of nested subsystems. In the center, the model accounts for individual characteristics, including personal dispositions, bioecological resources (e.g., ability, knowledge, experience, skills), values, and attitudes. The interrelated subsystems range in scale from the microsystem (e.g., home, school, neighborhood), to the exosystem (e.g., media, industry), and to the macrosystem (e.g., culture, policy). The chronosystem (not shown) accounts for influences over time or the influence of events occurring at a specific time. Behavior is shaped on these different "levels": from an individual's personal characteristics up to society's values and customs.

Interventions are designed to "pull levers" in multiple aspects of a system. Most notably, they target contextual factors, like the physical environment and policy. This is a break from traditional interventions that focus on individual knowledge, beliefs, attitudes, and behaviors. We know that in addition to these individual factors, the physical environment plays a crucial role in supporting healthy behaviors (Nisbet and Gick, 2008; Stokols, 1992b). Contemporary research and interventions related to physical activity, healthy eating, obesity, and chronic health conditions consider such factors as access to fresh foods or the presence of bike lanes. In other words, rather than asking, "how do we convince people in this neighborhood to walk more?," a public health leader might start by asking, "does this neighborhood have sidewalks?"

When it comes to people's role in product stewardship, this is the kind of thinking we need. Persuasive techniques won't convince someone to start walking or to return a product at the end of its life if they don't have access to safe places to walk or convenient drop-off locations. To minimize the environmental and health impacts of current technologies, ensure that newer resource-efficient technologies are adopted, and recover resources from consumers, it's essential that we better understand the relationship between human behavior and our environment. We can then take human behavior into account during product design and the design of product return or "take back" programs.

Recognizing and Avoiding Waste in Its Many Forms

We have more product choices today than ever, perhaps an infinite number of choices. Sitting at home we can purchase products from around the globe, typically with the promise of quick delivery. Amazon.com alone offers millions of products in every category imaginable, from food to furniture, fashion, cups, camping gear, cameras, cleaning products, and more. Everything we "need" at our fingertips. Yet what is the value in any of these products? The real value in products lies in the *function* they serve, or the need(s) that they fill (Brown et al., 2012). Anything that is not needed to derive that benefit—anything that can be eliminated while retaining the function of the product—is basically waste. While much of the focus here to this point has been on designing and enabling resource recovery, to the extent we can eliminate resource use in products, we can eliminate the need to design and put systems in place to recover resources.

Liquid laundry detergent was a relatively early target of modern corporate sustainability efforts, when companies began taking substantial steps to lessen their environmental impacts and resource use associated with their products, and provides a good illustration of how waste and resource use can be eliminated from product while retaining that product's value. In 2007 Walmart committed to selling only two-time concentrated detergent (Walmart, 2008). By doing so, the company estimated annual savings of 133 million gallons of water, 32 million pounds of plastic resin (in bottles), and 42 million pounds of cardboard (in packaging). Concentrated detergent also translates to fuel savings, as more units can be transported in less space. At the time, Walmart accounted for about 25 percent of liquid laundry detergent sales in the United States. The retailer drove manufacturers to reduce water, plastic, and cardboard use across the market. In April 2014, Walmart challenged manufacturers to reduce water in detergent another 25 percent (Procter & Gamble, 2014). Because of its strong influence on supply chains, commitments like this to only sell concentrated detergent had the effect of pushing others to make changes, even in the absence of regulation (Plambeck and Denend, 2008).

Just prior, Tide Pods hit the market in 2012. The single-dose laundry product made a splash in the sector, bringing in $500 million in revenue in its first year (Neff, 2012). Manufacturer Procter & Gamble marketed the pods based on cleaning efficacy, convenience, ease, and value. Though the product wasn't marketed in relation to company sustainability goals, it appeared to fit well into the company's overall sustainability strategy. The pods contained a highly concentrated 3-in-1 mix of detergent, stain remover, and brightener. Compared to standard liquid or even two-times concentrated liquid detergent, pods significantly reduced resource use associated with packaging and shipping (due to their smaller size and lighter weight). The Pod-encasing material was developed to dissolve in cold water, which supported the company's goals to increase the percentage of loads complete in cold water and save energy during use. The biggest impact of pods over the long term may come from the single-dose nature of the product itself. Pods relieved consumers from deciding how much detergent to use. The product is pre-measured, taking individual behavior out of the detergent "dosing" equation. This harkens back to setting a default option upon which consumers could rely.

Months after pods hit the market, competitors were apparently angered by the introduction of the single-dose packs and their impact on the market, saying pods were "killing the laundry detergent category" (Ziobro and Ng, 2013). Pods reduced the amount of product consumers used and even wasted. Pods had such an impact on the market that the overall sales of liquid laundry detergent in the United States actually dropped ostensibly due to reduced consumer waste. It turned out that manufacturers depended in part on consumers using more detergent than was actually needed to clean clothes.

The pod example reflects a way of thinking about resource efficiency and reducing waste. By reducing material use in the product and its packaging, shipping less water (weight), and changing the product's formulation, P&G was able to maintain the product's function while also saving significant resources.

Resource savings can be achieved in other ways. Some simply build off of existing models—like a washing machine that has an automatic detergent dispenser for "auto-dosing." Others, such as reusable polymer beads used during the wash cycle to reduce the need for both water and detergent, use a combination of existing and new technology.[1] Another way to look at resource savings in this context is to consider changing the design of what is being washed, rather than how it's washed. For example, with material advancements, a superfine wool fabric shirt requires washing much less frequently than its contemporaries.[2] Thus the structural design of the product (e.g., washing machine or superfine wool shirt) and product reformulation and packaging (e.g., detergent pods) offer significant design approaches that serve to advance resource conservation, along with other benefits (e.g., customer convenience and satisfaction).

When the now-established newcomer Method Products entered the laundry detergent market, CEO Adam Lowry emphasized the function of the products his company offered, saying, "We want to get paid for cleaning people's clothes, not for selling liquid."[3] Method Products introduced eight-times concentrated detergent in a pump-equipped bottle designed to make it easier to dispense the correct dose and use all of the detergent in the container.

New technologies might eliminate the need for cleaning products altogether. There are self-cleaning windows, self-cleaning cars, and other self-cleaning surfaces. Lowry predicted, "If you imagine what a fully sustainable laundry detergent looks like, it's probably not a liquid that gets consumed."[4]

As the examples above demonstrate, waste comes in many forms, like excess fuel used to transport non-concentrated liquid detergent that is primarily water; product residual that is difficult to extract from the bottom of a package; energy used to wash clothes in hot water; or excessive product use. Waste happens throughout the product life cycle, in varied ways, including the ancillary byproducts and materials disposed during production, distribution, and use (i.e., product and packaging remains). Waste is anything that doesn't contribute to a product's function and anything that doesn't deliver value or benefits. In fact, waste constitutes inefficiencies and economic loss.

1 Rethinking how we clean our clothes. Xeros Technology.
2 Why we wear merino wool. Wool&Prince.
3 The evolution of laundry detergent. Trellis.
4 Soap and glory: A peek behind Method's methods. https://trellis.net/article/soap-and-glory-peek-behind-methods-methods/.

Opportunities for Change

Companies face a host of reasons to decrease the toxicity of their products, minimize waste, and maximize resource recovery. Pressure from consumers, investors, and policymakers continues to grow. Potential liabilities related to pollution from PFAS and other chemicals loom. Increased awareness about environmental and human health has brought proper disposal into focus (see, e.g., Heacock et al., 2016). Recent events revealed the fragility of our supply chains and the possibility for material inputs suddenly to become difficult or impossible to obtain. Inflation eats away at margins. All of these factors and more make the need for businesses to focus on product stewardship abundantly clear. Businesses that don't minimize waste or recognize waste as a valuable resource are likely missing important opportunities.

Early in my career, I was discouraged by how many people it took to change a lightbulb. Now I see opportunity everywhere. There is no single solution, no simple method, and no single party responsible for product stewardship, although producers play a vital role in stewardship decision-making throughout a product's life cycle. Producer responsibility for their products begins with a product's initial conception and design and proceeds through material sourcing, manufacturing or assemblage, product distribution, customer support services, and product recovery (Fiksel, 2012). Recognizing that everything operates in a system with endless interactions helps us see this reality of shared responsibility and helps us think about the range of possibilities to achieve our goals—through changes in design, technology, policy, social norms, and our physical environment to name a few. By taking a long-term view and considering the entire life cycle of a product that includes recycling of materials back into the manufacturing process whenever possible to "close the loop" in a circular rather than linear process, decision makers can best identify opportunities to make changes within the system.

References

Akerlof, K. L., and Chris Kennedy. 2013. *Nudging toward a Healthy Environment: How Behavioral Change Research Can Inform Conservation.* Fairfax, VA: George Mason University.

Blackler, Alethea, Levi Swann, Marianella Chamorro-Koc, Wathsala Anupama Mohotti, Thirunavukarasu Balasubramaniam, and Richi Nayak. 2021. "Can We Define Design? Analyzing Twenty Years of Debate on a Large Email Discussion List." *She Ji: The Journal of Design, Economics, and Innovation* 7 (1): 41–70. https://doi.org/10.1016/j.sheji.2020.11.004.

Boks, Casper. 2012. "Design for Sustainable Behaviour Research Challenges." In *Design for Innovative Value towards a Sustainable Society*, edited by Mitsutaka Matsumoto, Yasushi Umeda, Keijiro Masui, and Shinichi Fukushige, 328–33. Dordrecht, Netherlands: Springer Netherlands.

Boks, Casper, and Johannes Daae. 2017. "Design for Sustainable Use Using Principles of Behaviour Change." In *Routledge Handbook of Sustainable Product Design*, edited by Jonathan Chapman, 316–34. New York: Routledge.

Booth, Sarah L., James F. Sallis, Cheryl Ritenbaugh, James O. Hill, Leann L. Birch, Lawrence D. Frank, Karen Glanz, David A. Himmelgreen, Michael Mudd, Barry M. Popkin, Karyl A. Rickard, Sachiko St. Jeor, Nicholas P. Hays. 2001. "Environmental and Societal Factors Affect Food Choice and Physical Activity: Rationale, Influences, and Leverage Points." *Nutrition Reviews* 59 (3): S21–36. https://doi.org/10.1111/j.1753-4887.2001.tb06983.x.

Bronfenbrenner, U., and P. Morris. 1998. "The Ecology of the Development Process." In *Handbook of Child Psychology*, edited by W. Damon and R. Lerner, 5th ed., 992–1028. New York: Wiley.

Brown, Howard J., Kristin Aldred Cheek, and Kathryn Lewis. 2012. *Naked Value: Six Things Every Business Leader Needs to Know about Resources, Innovation and Competition*. Edited by Mimi Egan. Illustrated edition. Guilford, CT: dMASS Media.

Campbell-Arvai, Victoria, Joseph Arvai, and Linda Kalof. 2014. "Motivating Sustainable Food Choices: The Role of Nudges, Value Orientation, and Information Provision." *Environment and Behavior* 46 (4): 453–75. https://doi.org/10.1177/0013916512469099.

Caterpillar, Inc. 2023. "About Caterpillar." https://www.caterpillar.com/en/company.html (accessed December 13, 2023).

Caterpillar, Inc. 2023. "Caterpillar's Circular Economy." www.caterpillar.com/en/company/sustainability/remanufacturing.html (accessed December 13, 2023).

Caterpillar, Inc. 2023. "Caterpillar's Remanufacturing Process." https://www.caterpillar.com/en/company/sustainability/remanufacturing/process.html (accessed December 13, 2023).

Croson, Rachel, and Nicolas Treich. 2014. "Behavioral Environmental Economics: Promises and Challenges." *Environmental and Resource Economics* 58 (3): 335–51. https://doi.org/10.1007/s10640-014-9783-y.

Dairy Business View. 2023. "New Study Validates Light Blocking Efforts." Dairyfoods.com. https://www.dairyfoods.com/articles/83251-new-study-validates-light-blocking-efforts (accessed October 30, 2023).

De Medeiros, Janine Fleith, Cecília Gravina Da Rocha, and José Luis Duarte Ribeiro. 2018. "Design for Sustainable Behavior (DfSB): Analysis of Existing Frameworks of Behavior Change Strategies, Experts' Assessment and Proposal for a Decision Support Diagram." *Journal of Cleaner Production* 188(July): 402–15. https://doi.org/10.1016/j.jclepro.2018.03.272.

"'Designing-in' Sustainable Behavior: A Nudge in the Right Direction." 2008. *Strategic Direction* 24 (11): 30–2. https://doi.org/10.1108/02580540810915039.

De Young, Raymond. 2000. "New Ways to Promote Proenvironmental Behavior: Expanding and Evaluating Motives for Environmentally Responsible Behavior." *Journal of Social Issues* 56 (3): 509–26. https://doi.org/10.1111/0022-4537.00181.

Fiksel, J. 2012. *Design for Environment: A Guide to Sustainable Product Development*, 2nd ed., New York: McGraw-Hill.

French, Simone A., Mary Story, and Robert W. Jeffery. 2001. "Environmental Influences on Eating and Physical Activity." *Annual Review of Public Health* 22 (1): 309–35. https://doi.org/10.1146/annurev.publhealth.22.1.309.

Gardner, G. T., and P. C. Stern. 2002. *Environmental roblems and Human Behavior.* Boston, MA: Pearson Custom Publishing.

Gifford, Robert. 2007. *Environmental Psychology: Principles and Practice.* Colville, WA: Optimal Books.

Heacock, Michelle, Carol Bain Kelly, Kwadwo Ansong Asante, Linda S. Birnbaum, Åke Lennart Bergman, Marie-Noel Bruné, Irena Buka, David O. Carpenter, Aimin Chen, Xia Huo, Mostafa Kamel, Philip J. Landrigan, Federico Magalini, Fernando DiazBarriga, Maria Neira, Magdy Omar,Antonio Pascale, Mathuros Ruchirawat, Leith Sly, Peter D. Sly, MartinVan den Berg, and William A. Suk. 2016. "E-Waste and Harm to Vulnerable Populations: A Growing Global Problem." *Environmental Health Perspectives* 124 (5): 550–5. https://doi.org/10.1289/ehp.1509699.

Hood. 2023. "The Hood lightblock bottle." https://hood.com/lightblock-bottle (accessed October 30, 2023).

Kaplan, Stephen. 2000. "New Ways to Promote Proenvironmental Behavior: Human Nature and Environmentally Responsible Behavior." *Journal of Social Issues* 56 (3): 491–508. https://doi.org/10.1111/0022-4537.00180.

Kollmuss, Anja, and Julian Agyeman. 2002. "Mind the Gap: Why Do People Act Environmentally and What Are the Barriers to pro-Environmental Behavior?" *Environmental Education Research* 8 (3): 239–60. https://doi.org/10.1080/13504620220145401.

Kurz, Tim. 2002. "The Psychology of Environmentally Sustainable Behavior: Fitting Together Pieces of the Puzzle." *Analyses of Social Issues and Public Policy* 2 (1): 257–78. https://doi.org/10.1111/j.1530-2415.2002.00041.x.

Lilley, Debra. 2009. "Design for Sustainable Behaviour: Strategies and Perceptions." *Design Studies* 30 (6): 704–20. https://doi.org/10.1016/j.destud.2009.05.001.

Lockton, Dan, David Harrison, and Neville Stanton. 2008. "Making the User More Efficient: Design for Sustainable Behaviour." *International Journal of Sustainable Engineering* 1 (1): 3–8. https://doi.org/10.1080/19397030802131068.

McLeroy, Kenneth R., Daniel Bibeau, Allan Steckler, and Karen Glanz. 1988. "An Ecological Perspective on Health Promotion Programs." *Health Education and Behavior* 15 (4): 351–77. https://doi.org/10.1177/109019818801500401.

Neff, Jack. 2012. "Tide Pods Winning $7 Billion Detergent Wars by Redefining Value: How Pods Defied the Odds in a Crowded Category." *Ad Age*. December 18, 2012.

Nisbet, Elizabeth K. L., and Mary L. Gick. 2008. "Can Health Psychology Help the Planet? Applying Theory and Models of Health Behaviour to Environmental Actions." *Canadian Psychology* 49 (4): 296–303.

Pichert, Daniel, and Konstantinos V. Katsikopoulos. 2008. "Green Defaults: Information Presentation and Pro-environmental Behaviour." *Journal of Environmental Psychology* 28 (1): 63–73. https://doi.org/10.1016/j.jenvp.2007.09.004.

Plambeck, Erica, and Lyn Denend. 2008. "The Greening of Walmart." *Stanford Social Innovation Review*, 2008. https://doi.org/10.48558/zv0b-f456

Procter & Gamble. 2014. "Tide® Challenges Americans to Switch to Cold Water During Earth Week."

Razani, Nooshin, and June Tester. 2010. "Childhood Obesity and the Built Environment." *Pediatric Annals* 39 (3): 133–9. https://doi.org/10.3928/00904481-20100223-04

Reisch, Lucia A., and Cass R. Sunstein. 2014. "Redesigning Cockpits." *Journal of Consumer Policy* 37 (3): 333–9. https://doi.org/10.1007/s10603-014-9267-z

Sallis, James F. 1998. "Environmental and Policy Interventions to Promote Physical Activity." *American Journal of Preventive Medicine* 15 (4): 379–97. https://doi.org/10.1016/S0749-3797(98)00076-2.

Schultz, P. Wesley. 2014. "Strategies for Promoting Proenvironmental Behavior: Lots of Tools but Few Instructions." *European Psychologist*, Special Section: Environmental Conservation., 19 (2): 107–17. https://doi.org/10.1027/1016-9040/a000163.

Selvefors, Anneli, Karin Blindh Pedersen, and Ulrike Rahe. 2011. "Design for Sustainable Consumption Behaviour: Systematising the Use of Behavioural Intervention Strategies." In *Proceedings of the 2011 Conference on Designing Pleasurable Products and Interfaces*, 3:1–3:8. DPPI '11. New York: ACM. https://doi.org/10.1145/2347504.2347508.

Smith, Susan. "Opaque Milk Jugs Contain Recycling Issues." May 25, 1998. *Plastics News*. https://www.plasticsnews.com/article/19980525/NEWS/305259974/opaque-milk-jugs-contain-recycling-issues (accessed October 30, 2023).

Steg, Linda, and Charles Vlek. 2009. "Encouraging Pro-environmental Behaviour: An Integrative Review and Research Agenda." *Journal of Environmental Psychology*, 29 (3): 309–17. https://doi.org/10.1016/j.jenvp.2008.10.004.

Stern, Paul C. 2000. "Toward a Coherent Theory of Environmentally Significant Behavior." *Journal of Social Issues* 56 (3): 407–24. https://doi.org/10.1111/0022-4537.00175

Stokols, Daniel. 1992a. "Environmental Quality, Human Development, and Health: An Ecological View." *Journal of Applied Developmental Psychology* 13 (2): 121–4. https://doi.org/10.1016/0193-3973(92)90021-9.

Stokols, Daniel. 1992b. "Establishing and Maintaining Healthy Environments: Toward a Social Ecology of Health Promotion." *American Psychologist* 47 (1): 6–22. https://doi.org/10.1037/0003-066X.47.1.6.

Story, Mary, Karen M. Kaphingst, Ramona Robinson-O'Brien, and Karen Glanz. 2008. "Creating Healthy Food and Eating Environments: Policy and Environmental Approaches." *Annual Review of Public Health* 29 (1): 253–72. https://doi.org/10.1146/annurev.publhealth.29.020907.090926.

Sunstein, Cass R., and Lucia A. Reisch. 2014. "Automatically Green: Behavioral Economics and Environmental Protection." *Harvard Environmental Law Review* 38: 127–58. http://dx.doi.org/10.2139/ssrn.2245657

Tang, Tang, and Tracy A. Bhamra. 2009. "Understanding Consumer Behaviour to Reduce Environmental Impacts through Sustainable Product Design." In *Undisciplined! Design Research Society Conference 2008*, 1–15. Sheffield Hallam University, Sheffield, UK.

Thaler, Richard H., and Cass R. Sunstein. 2003. "Libertarian Paternalism." *The American Economic Review* 93 (2): 175.

Walmart Stores, Inc. 2008. "Walmart Completes Goal to Sell Only Concentrated Liquid Laundry Detergent." May 29, 2008.

Wells, Nancy M., Susan P. Ashdown, Elizabeth H. S. Davies, F. D. Cowett, and Yizhao Yang. 2007. "Environment, Design, and Obesity: Opportunities for Interdisciplinary Collaborative Research." *Environment and Behavior* 39 (1): 6–33. https://doi.org/10.1177/0013916506295570.

Winter, Deborah Du Nann, and Susan M Koger. 2004. *The Psychology of Environmental Problems*. Mahwah, NJ: Lawrence Erlbaum.

Ziobro, Paul, and Serena Ng. 2013. "Is Innovation Killing the Soap Business?" *Wall Street Journal*. April 3, 2013.

2

The Public Sector Takes on Latex Paint Recycling and Gains a Partner

Jim Quinn
Solid Waste Resilience Manager, Metro, Portland, Oregon

In its broadest sense, stewardship is about taking responsibility and providing careful management of something, usually for the greater good. In a best-case scenario, companies that produce goods are the stewards of their products throughout the product life cycle. After all, they are the most knowledgeable about their products and they reap the benefits from their sale. Government also steps in as needed, and serves in a stewardship role for shared product responsibility.

This chapter describes the evolution of paint stewardship and an industry that ultimately partnered with the government—in particular, with a local government agency that noticed a problem and saw it as an opportunity. In fact, the architectural paint industry in the United States has risen to the status of a product steward exemplar, with growth in Extended Producer Responsibility (EPR) for paint. This chapter provides background on latex paint generation, issues, and management. It details the experience (over three decades) of the vital interplay of government and industry during that time, with good resolution and anticipated progress in the future. The chapter concludes with some lessons learned, many of which can also be applied to products other than paint.

Background: post-consumer Latex Paint Problems and Solutions

In the mid-twentieth century, architectural paint became widely available as a consumer product, tapping into the growing do-it-yourself movement of the time. Architectural paint is the kind used to decorate and protect interior and exterior surfaces of homes and other buildings, and it includes both oil-based and latex paint (also known as water-based paint). Latex paint provided advantages over traditional oil-based paint, including ease of cleanup and a less overpowering odor. Still, many painters remained loyal to oil-based paint believing it provided superior performance. That is, until the 1990s, when concerns surfaced about air quality and health impacts of the solvents found in oil-based paint. These concerns, along with significant improvements in the quality of latex paint formulations, resulted in a major market shift with oil-based paint now mostly relegated to a few specialty applications.

Latex paint has a much better safety profile than oil-based paint because it is water-based and does not have the health and environmental impacts or fire risks posed by the solvents found in oil-based paint. Latex paint is not, however, trouble-free. Trouble rises from its popularity: a surprisingly large volume of latex paint waste is generated in the United States each year. Best estimates are that about 10 percent of all paint that is purchased becomes unwanted leftovers.[1] Based on current sales, that translates to an estimated 80 million gallons of leftover paint generated annually in the United States.

So what options are there for dealing with all of this post-consumer paint? Government waste authorities generally do not allow latex paint to be disposed of in the trash due in large part to its liquid nature and the sheer volume of paint in the marketplace. There are several ways that latex paint can be rendered non-liquid (described later) and thus allowable in trash in most locales. However, it should be noted that some jurisdictions consider latex paint to be hazardous under local regulation, especially older paint that contains now-discontinued ingredients, and thus disposal in trash is not allowed, even when it is solidified.

1 US EPA, *Quantifying the Disposal of Post-Consumer Architectural Paint*. 2007. EPA. https://archive.epa.gov/sectors/web/pdf/paint_quantity_report.pdf (accessed August 22, 2022).

Consumers' options for unwanted latex paint when no collection programs are available include the following:

- *Give it away*—If the paint is still good, find someone else who can make use of it.
- *Use it up*—Apply it to a garage or utility room wall.
- *Dry it out*—Pop the lids and leave cans with partial content in a secure place to dry out over time, or even pour it onto newspaper or thick millimeter plastic to solidify it to throw away.
- *Add to it*—Available at hardware stores, various additives can be mixed into paint to make it solid. Materials like sawdust, kitty litter, or shredded newspaper can also be used.

These options are not without challenges. They can be labor-intensive and messy. Many people do not have space in their home to dry out or solidify paint. With a typical American home having several cans stockpiled, many residents just end up leaving the paint in storage, often passing on the problem when they move out. In a 2004 EPA study, leftover latex paint brought to collection programs averaged seven years in age, with some cans as old as twenty-five years.[2] Clearly there is a lot of paint stored in garages and basements everywhere.

What about pouring it down the drain? Highly diluted latex paint from washing out brushes is generally acceptable for municipal sewage treatment systems, though even diluted, it can be harmful in septic systems. Pouring undiluted latex paint down a household drain is never a good idea as it can potentially clog drains and is harmful to the beneficial bacteria in wastewater treatment plants. Poured into a storm drain or anywhere outdoors, it will eventually find its way to a local stream where it is harmful to aquatic life.

With the advent of household hazardous waste (HHW) collection programs in the early 1980s, a better option became available. HHW collection grew steadily through the decades following, and today there are thousands of communities in the United States that have access to collections. Most HHW programs accept latex paint alongside other problematic household wastes. In

[2] "Quantifying the Disposal of Post-Consumer Architectural Paint." 2007. EPA. https://archive.epa.gov/sectors/web/pdf/paint_quantity_report.pdf.

fact, paint (both latex and oil-based) often comprises 50 percent or more of HHW collected. Some communities conduct targeted collections of just a few more common residential wastes, such as collecting just antifreeze, batteries, oil, and paint, sometimes called "ABOPs."

Some HHW program managers have chosen not to accept latex paint, focusing their limited resources on the more hazardous household products. Taking latex paint out of the mix of acceptable materials at HHW collections can however have a boomerang effect. That's because the resident's desire to clear away paint cans often expands to include other HHW (e.g., cleaners, pesticides), further motivating program participation. Thus, the effectiveness of the collection program is reduced when residents choose to forgo participation thereby leaving the more hazardous stuff in storage as well, because they cannot also get rid of their paint.

As early as the 1990s, some jurisdictions experimented with paint-only collections at retail paint stores to provide a more convenient option than HHW collection offerings. While promising, limited funding prevented expansion of this option, until the advent of paint stewardship as detailed later in the section on partnering with the paint industry.

So what do collection programs do with the latex paint they collect? Many recycle it; some methods of which are presented in the following sections. However, there is always a significant fraction of collected latex paint that is not "up to snuff" for recycling and thus requires proper disposal. Some programs have opted to dispose of all collected latex paint, finding it their lowest cost option. Some solidify it for solid waste disposal, much like residents can do, but on a larger scale. This can be accomplished by mixing it with solidification agents in drums or larger containers, even in a large recessed paved area using a backhoe. Besides the solidification agents available to residents, various industrial products and by-products have been used in bulk latex paint solidification.

Other options that have been used include mixing it in with wastes with a higher energy value destined for energy recovery facilities where it is burned as a fuel, or even finding a beneficial use for latex paint waste in the production of non-paint products. This has included mixing it with concrete for nonstructural uses, such as jersey barriers, paving stones, or parking blocks (used at the head of spaces in parking lots). A few programs have used it as part of alternative daily cover material for solid waste landfills.

All of these disposal methods come with added costs, for materials, labor, transportation, and so on. And when disposal processes are used for paint that is still good, they can be viewed as wasting a valuable recyclable resource. Recycling does come with costs as well, but can provide significant environmental benefits, and there can be revenue from selling the resulting recycled paint products. (Further discussion of costs vs. benefits is found in section "The State of the Art Today.")

According to the Product Stewardship Institute (PSI),

> About 10 percent of all household latex and oil-based paint goes unused in the United States—that's about 80 million gallons each year. When dumped in the trash or down the drain, unused paint can contaminate our environment with volatile organic compounds, fungicides, and (in the case of very old paint) hazardous metals such as mercury, lead, cadmium, and hexavalent chromium. Although most leftover paint can be easily reused, recycled into new paint, or repurposed into other recycled products, much of it is trashed because people don't know where to take it and often have no other good options. Paint is the most expensive product for local household hazardous waste programs to manage, costing governments and taxpayers across the country millions of dollars.[3]

As HHW collection programs grew and matured in the 1980s, some managers turned their attention to improving the environmental impacts of the waste disposition methods they employed. By the late 1980s a few pioneers had tried their hand at recycling latex paint. This included local governments operating HHW collection programs, companies that made paint, and waste processing and disposal firms. Many of these forays into paint recycling were short-lived and the results were decidedly mixed; however, a lot of learning took place, and paint recycling was born.

[3] Paint. Product Stewardship Institute. https://www.productstewardship.us/page/Paint (accessed August 13, 2022).

Terminology: Reuse, Reprocessing, Reblending, Recycling

There are several terms that have been used in connection with reuse and recycling of latex paint; unfortunately there is not always consistency in the meaning and usage of these terms. For the purposes of this chapter, some key terms and definitions follow.

Reuse typically refers to finding users for unwanted paint still in the containers that it was originally purchased in. Reuse in this sense can be a valuable strategy for dealing with unwanted paint. *Paint Reuse: A Manual for Developing Effective Programs*[4] states,

> Convenient paint reuse programs can reduce the environmental and financial costs of left-over paint and decrease waste while providing a valuable service to the community. Reuse is considered one of the most environmentally preferable waste management strategies, topped only by source reduction on the waste management hierarchy. Paint reuse programs provide a low-cost, local alternative to recycling, while establishing infrastructure that may be eligible for compensation under a future producer-funded paint stewardship program.

However, there are limitations to this strategy and it can conflict with the goals of paint recycling programs, as it can result in "cherry-picking" some of the most desirable paint, leaving less valuable paint available for further recycling. This chapter does not address reuse further; for those interested, the manual mentioned above is an excellent resource.

The term *recycling* has been used in various ways in relation to paint, but generally refers to removing the paint from its original container and processing it in some way, and ultimately having it end up in a new container for distribution. Procedures can vary tremendously. On the easy end of the spectrum, paint is processed by pouring together the contents of a few containers of leftover paint into a new container, such as a 5-gallon pail, mixing

[4] *Paint Reuse: A Manual for Developing Effective Programs*. 2015. New York Product Stewardship Council (NYPSC) and the Product Stewardship Institute, Inc. (PSI). https://productstewardship.us/press_releases/product-stewardship-institute-launches-best-practices-guide-for-paint-reuse-in-new-york/ (accessed May 29, 2022).

it up, putting a lid on it and calling it good. Often, like-colors are sorted and poured together. This low-tech approach has the advantage of being quick and easy and requires minimal equipment.

It is important to examine the quality of each individual container being poured out, as a number of things can compromise the quality of the paint: freezing, bacterial spoilage, drying out, clumping, and foreign debris all being common problems. A visual check is a good low-tech option. The contents of each container can be stirred, looking for signs of anything that may compromise paint usability. Another approach is to pour the paint through screening material, such as a metal screen or paint filtering cloth.

Sometimes the term *reblending* is used to describe this simple approach. Reblending can be scaled up to 55-gallon drums or larger containers, with the consolidated paint then dispensed into consumer-sized containers. The term *consolidated* has sometimes been used instead of *reblended.*

The paint product that results from low-tech reblending can be very inconsistent in quality. The final color will be unpredictable, bacterial spoilage may be an issue, and really nothing about the product is consistent from batch to batch. Color consistency can be improved by going to larger batches, and other improvements can be implemented to improve product quality,

However, reblending is at the low end of the paint recycling continuum. The value of reblended product is generally considered low. Some programs give away reblended paint, some charge a nominal amount per gallon. To produce paint with more value a lot more needs to be done. Improvements in color consistency, filtering, addition of various additives, and other process improvements greatly increase the consistency and value of the resulting product.

Sometimes the term *reprocessing* is used to describe processes that incorporate these quality-enhancing features. Additional detail on these process improvements is listed below. Another approach to improving quality is to use recycled content as only a portion of product formulation, using virgin raw materials for the remainder.

US EPA Definitions

Reprocessed paint is post-consumer latex paint that has been sorted by a variety of characteristics including type (i.e., interior or exterior), light and dark colors, and finish (e.g., high-gloss or flat). Reprocessed paint is available in various colors and is suitable for both interior and exterior applications.
Consolidated paint consists of post-consumer latex paint with similar characteristics (e.g., type, color family, and finish) that is consolidated at the point of collection. Consolidated paint is typically used for exterior applications or as an undercoat.

Source: Comprehensive Procurement Guidelines for Construction Projects. US Environmental Protection Agency. https://www.epa.gov/smm/comprehensive-procurement-guidelines-construction-products#05 (accessed April 10, 2022).

The blossoming of latex paint as a favored product for both do-it-yourselfers and professional painters, and the resulting volume of post-consumer leftover paint, clearly posed challenges early on with limited available management options. But it also presented opportunities, as shown by some of the earliest forays into collection and recycling. This was the landscape in the early 1990s when the regional government in Portland, Oregon, turned its attention to addressing the problem, and embracing the opportunity.

Portland Metro's Paint Recycling History

Metro is the regional government in the Portland, Oregon, area, with a variety of governmental responsibilities including planning and overseeing the region's solid waste system. These include operating two municipal solid waste transfer stations, both of which have HHW collection facilities. The Metro region is a mix of urban and suburban communities and includes the City of Portland and twenty-three suburban cities, with a total population of about 1.6 million.

Metro's approach to solid waste management often includes leading-edge strategies. As one example, in the early 1980s local governments were beginning to grapple with the environmental and public health risks of HHW. With no one else stepping up to address the problem, some local governments established education and collection programs to deal with HHW. Metro conducted periodic HHW collection (single-day) events starting in 1986, and soon began planning to open permanent HHW collection facilities in the region that would greatly extend the number of open-days and convenience for Portland residents.

The Early Days of Low-Tech Recycling

Metro's planning for a permanent HHW program focused on the opportunities provided by having custom-designed facilities staffed with permanent Metro employees. Metro was well aware that latex paint would be a significant percentage of the waste received when the first facility opened its doors and began planning for receiving and managing this volume of paint.

I was hired by Metro in 1991 to implement the HHW program startup, and on my first day of work my boss tasked me to figure out what we were going to do with all that latex paint. From the beginning Metro was set on emphasizing recycling and recovery of HHW wherever possible, so our focus was on developing a recycling process for the latex paint we were to receive. We looked at two municipal paint recycling programs we knew of that had pioneered the way, in order to understand what had worked and what hadn't worked. Metro also conducted a paint-only collection event to learn more about the reuse option. We then sketched out our game plan, procured supplies, and hired staff, and when Metro's first HHW facility at the Metro South Transfer Station in Oregon City opened its doors in February of 1992; we were ready to recycle paint.

At that time our process was very much on the low- tech end of the continuum. The space we used was a large concrete structure adjacent to the transfer station that had previously functioned as a solid waste loading bay—or the "tunnel" as it is affectionately known. The system was based on 55-gallon steel drums with wire mesh placed over the top, with drum mixing accomplished manually using a boat paddle. Incoming 1- and 5-gallon containers of latex paint were opened and stirred, initially with a small wooden paint-mixing stick. A basic quality sort was done at this point and anything obviously wrong with the paint meant it would not go to the recycling drums. Next was a rough color sort, with like colors being poured with like, through the wire mesh, into the drum and stirred. No separation of interior and exterior varieties or of sheens was done. That was it. These drums of paint were reasonably good quality, though color certainly varied from drum to drum. We offered the drums at no cost to anyone who could use them. Even so it took quite a bit of staff time to round up willing customers who were able to utilize all of the drums of paint we produced.

Initially the recycle rate was rather low, around 38 percent, leaving a lot of nonrecyclable paint to manage. We tried a variety of methods to deal with this. Since latex paint is not considered hazardous waste under Oregon or federal rules, disposal was primarily a matter of efficiently turning large volumes of paint into a solid. We experimented with some on-site methods, but ultimately settled on paying our HHW disposal contractor to take it and solidify it in bulk using a backhoe to mix it with absorbent agents at their facility.

From these early days to today, Metro has been engaged in a continual process of improving the paint recycling process in many different ways. An early improvement was the switch from stick stirring of opened incoming cans and pails to use of a mixer mounted on a powered hand-operated drill. It turns out that a lot of the paint that was rejected for recycling was actually perfectly good, but it took a vigorous stirring process to confirm this. The recycle rate immediately increased to nearly 50 percent at that time, with additional increases to come in the future.

Experiments with making targeted colors was another early improvement. Of course, paint is all about color. Early on, it became apparent that customers preferred some colors over others. We were always tweaking colors to try to maximize the desirability of each drum of paint.

First Partnership with the Private Sector

Early on in our low-tech days it became apparent that there was a limit to the types of customers who are able to utilize paint in drums. This led to our first foray in partnering with a local paint manufacturer that utilized a portion of our reblended paint as a feedstock in their manufacture of paint for retail sale.

In standard government format, we conducted a Request for Proposals (RFP) process, with all of the requisite contract terms and legal language, and sent it out to a list of regional paint manufacturers. Miller Paint, a leading regional paint manufacturer and retailer, sent a letter expressing some interest in possibly working with us but rejecting the various terms and conditions a government RFP process required.

A more favorable response came from a small local manufacturer called Rasmussen Paint (no longer in business). Under the resulting agreement, we provided them with paint in 55-gallon drums, at no cost, sorted during our consolidation process into three varieties: an interior light, an exterior light, and an interior-exterior dark. Rasmussen then filtered the paint and mixed in additives and raw materials as they saw fit for quality control, to produce paint for sale, labeled as recycled paint. This was a fruitful partnership, providing an outlet for about half of the paint we produced for a couple of years.

Metro's First Custom-Designed Paint Recycling Facility

Eventually it became apparent that Rasmussen was unable to incorporate a significant portion of our paint into their product. Also, with increasing volumes of paint coming into our HHW program, we had experienced challenges with ergonomic injuries to the workers doing the processing: musculoskeletal strains and sprains to workers' backs, shoulders, and the like, brought on by the hard work of handling paint containers. Our focus shifted to designing a new recycling process from scratch, geared toward producing paint in 5-gallon pails, which we believed we could sell. Ergonomic considerations were also foremost in the design. Staff members who had hands-on experience with the low-tech process collaborated with Metro engineering staff to develop the design and procure the necessary equipment, much of it custom fabricated. The layout they developed still forms the heart of the process we use today, more than twenty-five years later. A new building was constructed on the site to house the operation, which in 1999 moved out of the tunnel that had housed it for the first several years.

The improved process was based on a number of concepts:

- an elevated mezzanine where paint cans are opened and poured into sinks, which lead to bulk containers below;
- custom fabricated stainless steel tables, with carefully calculated table heights and workstation distances to minimize ergonomic strain;
- IBCs: commonly available pallet-mounted polyethylene containers, around 250–300 gallons in capacity, are known as IBCs (intermediate bulk containers), which provide advantages over 55-gallon drums;
- mixing: the larger volume of paint in an IBC is more difficult to mix than a 55-gallon drum, requiring powerful industrial mixers to blend the paint; and
- packaging into 5-gallon containers: initially we tried cleaning out and reusing the incoming 5-gallon pails from incoming waste paint but found this presented too many challenges and procured new pails.

The Launch of MetroPaint

The new processing facility opened up a major new opportunity to sell 5-gallon pails of paint to the public. Producing batches at the IBC scale, we found that some consistency in color from batch to batch started to emerge. For example, light blues may go into one tank, dark blues into another, and the averaging that takes place in the blending process results in some similarity between subsequent batches. To facilitate sales, Metro's Public Affairs Department helped develop labeling for the product (Figure 2.1). We developed a material safety data sheet (MSDS)—a legal requirement for a product like paint—and began selling the paint at the Metro South Transfer Station site.

Selling a retail product is not something that government agencies typically get into, so a number of new administrative and financial processes needed to be developed. Fortunately, Metro has a culture of innovation and provided the support for blazing new trails. We learned a lot as we went; see the detailed discussion of the marketing challenges on the following pages and our strategies for overcoming them.

Before long we recognized that offering only 5-gallon pails was a significant limitation to our potential market, and in order to increase sales we developed a process for dispensing MetroPaint into 1-gallon cans.

Figure 2.1 MetroPaint labels over the years.

Source: Jim Quinn.

While the new processing facility was a quantum leap in quality and marketability, there were many innovations and improvements still to come. Over the years key process improvements have included the following:

- industrial paint filtration equipment,
- the addition of biocides to minimize odor and provide resistance to microbial growth,
- a variety of quality control tests and adjustments,
- consistent color-matching processes,
- larger tanks and improved mixing and pumping equipment, and
- reduced water usage.

We also increased the volume that was processed, particularly when we began taking in paint from neighboring jurisdictions' paint collection programs.

Partnering with the Paint Industry

After Metro's partnership with Rasmussen Paint ended, we operated successfully for several years without any connection to the paint industry.

As a government agency, we provided a convenient way for the region's residents to get rid of unwanted paint, while making and selling an affordable, environmentally preferable, recycled paint product.

Around the country, though, many still did not have a good way to get rid of their unwanted paint. The number of government-operated HHW programs grew steadily for a couple of decades but largely plateaued after the 1990s. Many portions of the country did not have any HHW collection programs (and still don't to this day). Some HHW programs around the country did develop in-house latex paint recycling processes, generally of the low-tech variety, with a few aspiring to make a more valuable product that could bring in some revenue to their program.

For a long time, the paint industry was generally not supportive of the concept of paint recycling. And they did not have any interest in being part of the solution to dealing with all of the leftover paint piling up in communities. I attended a government-sponsored meeting about paint recycling in the early 1990s and heard industry representatives talking about how easy it was for homeowners to solidify latex paint for trash disposal, how untenable the idea of recycling paint was due to possible contaminants that people could add to their paint cans, and how once they sold the paint it was no longer their responsibility.

The mid-1990s saw the beginning of conversations about manufacturers stepping up to be responsible in some way for the proper management of their products at the end of the product's life cycle. While terminology has evolved, the term *product stewardship* was used to describe this concept, and later the term *Extended Producer Responsibility* (EPR) was adopted. In Europe, this concept was applied to packaging waste as early as 1991. In the western hemisphere, British Columbia, Canada, was an early pioneer, establishing a product stewardship program for paint in 1994. When we heard about this concept it seemed like a pie-in-the-sky idea that we could only dream about here on the other side of the border.

In Oregon, the Department of Environmental Quality (DEQ) wanted to explore the concept of collecting leftover paint at retail locations. In 1997, a pilot program known as PaintSmart was launched to test the collection of paint at paint retail stores. Eleven stores statewide participated in this voluntary pilot program, accepting leftover paint from residents for recycling and safe disposal. Paint stores provided the space and and time for staff to accept paint from customers; state and local governments paid other costs. The program was discontinued in 1998, but in the Portland area, one Miller Paint store continued offering the service, with Metro picking up and recycling the paint; this store continues to be a paint collection point today.

The Paint Dialogue Launches

Despite the seeming unlikelihood of adopting the British Columbia strategy here in the United States, change was on the horizon. The PSI, a nonprofit organization based in Massachusetts, launched in 2000 with paint as an early focus product. In 2001, PSI started developing background information about the leftover paint problem and in 2002 reached out to the paint industry to see if they might be willing to discuss this concept of product stewardship. In 2003, PSI launched a multi-stakeholder dialogue process, which led to several years of engagement and exploration with the industry and other stakeholders, eventually leading to groundbreaking change.[5]

A key element of the dialogue process was the development of a variety of reports and studies addressing various aspects of a potential paint stewardship program, and of paint recycling, including the following:

EDUCATION

1. Leftover Paint Management and Guidance for Consumers
2. Public Education Survey and Analysis

INFRASTRUCTURE

3. Paint Reuse Primer
4. National Infrastructure Model
5. Infrastructure Cost Analysis

MARKETS

6. Market Development Strategy for PPSI
7. Recycled Paint Marketing Guide for Distributors
8. Recycled Paint Certification System

[5] Paint. Product Stewardship Institute, Inc. 2023. https://productstewardship.us/products/paint.

OTHER PROJECTS

9. Health-Safety-Environmental and Regulatory Issues
10. Financing System Research and Model Development
11. Life Cycle Balance of Costs and Benefits

Metro played an important role in the engagement with industry on these projects. Many in the paint industry were initially skeptical about the possibility of recycling latex paint. For some, it was a concern about bringing leftover paint into their production facilities for fear of microbial contamination. This is a valid concern. Think about painters dipping their brush into the can, painting their wall, then putting their brush back in the can. This can certainly introduce contaminants into paint, which could easily contaminate manufacturers' virgin raw materials. There was also a reluctance to accept the idea that paint made from a wide array of leftover brands mixed together could be both good quality and safe. My thought was that manufacturers spend so much time perfecting and fine-tuning their own formulations that this just sounded fundamentally wrong to them.

The existence of Metro's program, in which we encountered (and solved) many of the technical problems, served as a powerful argument for the viability of recycling. Along with other recyclers involved in the dialogue, we addressed concerns about contaminants in recycled paint, the viability of markets, and other real-life aspects of a successful latex paint recycling operation.

The paint dialogue eventually resulted in a sea change in the paint industry. In 2007 a memorandum of understanding among the various parties solidified industry's commitment to developing an industry-operated system for collection and management of leftover paint. Though no one on the industry side has come out and said as much, many believe that their decision to take on this responsibility was motivated by the desire to be seen as a good corporate citizen, in hopes of overcoming long-standing image problems resulting from historical lead-based paint exposures.

It took a couple more years to work through the legislative process, taking place at the state level, but in 2009 the first bill resulting from the dialogue was passed. It turns out that this bill passage occurred in Oregon. The stakeholder group had attempted to pass a bill in Minnesota, but political issues prevented that on two attempts. Because of the participation and support shown by Metro and other Oregon stakeholders during the dialogue process, the legislative baton was passed on to Oregon. With full industry support, the first paint stewardship bill in the United States was passed in the Oregon 2009 legislative session, making us the laboratory state for paint product stewardship in this country. Metro was eager to play a key role.

The Role of Private Sector Paint Recyclers

In the early days of experimental paint recycling programs, starting around 1989, a variety of organizations tried their hand at recycling latex paint, including local governments, paint companies, and waste management firms. The first private sector paint recycling company that was able to sustain operations over time was Amazon Paint, established in 1992. A handful of other companies have joined the ranks over the years, though this service is still not available in many regions of the country.

Some of these recyclers participated in the PSI paint dialogue process. A couple were stand-alone recyclers, others were affiliated with the paint manufacturing industry. Their role in the dialogue process was important. For example, they addressed the concern about microbial contamination spreading to virgin raw materials by operating stand-alone paint recycling facilities that did not involve any virgin paint manufacturing. However, in many cases they were reluctant to share technical information that could give their competitors an advantage. This of course was not a concern for Metro as a government agency.

Revisiting Local Partnerships with Industry

As the national paint dialogue process played out, Metro decided to make another attempt at collaborating with local paint companies. The most likely candidate was Oregon-based Miller Paint. Although they had rejected our overtures years earlier, by the mid-2000s, Miller Paint had new leadership in place. We learned that they were supportive of paint recycling in general and saw Metro's recycling program as a great solution to the leftover paint problem.

With this foundation we entered negotiations with them, which eventually led to an agreement for them to carry MetroPaint in Miller Paint stores. More details on the marketing benefits of this relationship are detailed later.

Over the years the partnership has encompassed other elements, including technical assistance of various kinds, procurement and production of containers for packaging MetroPaint, and more. One other local paint manufacturer, Rodda Paint, has also provided a variety of technical assistance to us, as we developed some of our more advanced laboratory testing and additive protocols.

PaintCare Launches in Oregon

With the passage of the Oregon bill mandating a paint collection system in the state in 2009, a lot of work needed to be done to get the program up and running by the mandated start date, July 1, 2010. The American Coatings Association (ACA), the paint industry organization responsible for setting it up, hired a Canadian organization, Product Care, to plan and implement operations. Product Care had experience operating product stewardship systems across Canada, including the British Columbia paint stewardship program.

To create the system, Product Care needed to recruit collection points, transporters, and recycling or disposal contractors. In a unique twist, they issued a request for proposals (RFP), a tool typically used by government agencies to procure goods and services. As an aspiring contractor to Product Care, Metro needed to develop a proposal in response to the RFP. Being in this position was a new challenge for Metro. Determining all of our costs for collection, transport, and recycling and setting the appropriate pricing to charge Product Care presented a particular challenge. Metro's proposal was that we would recycle not only the paint collected here in the region but also the paint brought to us from all around the state under the program. Product Care accepted our proposal, and a contract was negotiated. Shortly after 9:00 a.m. on July 1, 2010, the first can of paint collected under a US product stewardship system was unloaded from a customer's car at the Metro South HHW facility in Oregon City.

A requirement of the Oregon bill was that the producer-run system provide education to residents regarding buying the right amount of paint in the first place, using up paint when possible, and bringing leftover paint to the collection points in the newly established statewide infrastructure. Over the course of the dialogue process ACA had made it clear that they would not promote any particular brand of recycled paint, due to concerns of their paint-retailing members. Consequently the public education materials for the Oregon program originally did not direct consumers to MetroPaint. However, all of the promotional materials did describe bringing their paint into the program as recycling it, so that support for recycling was built in from the beginning. Eventually the industry program, which came to be known as PaintCare, would evolve, and would tell consumers where they could purchase recycled paint, including MetroPaint.

Highlights of the Oregon Paint Product Stewardship Program

As of 2020, following are the highlights of the Oregon Paint Stewardship Program:

- The program convenience goal of a permanent collection site for every 30,000 residents has been exceeded with 177 drop-off sites, including 28 HHW or Solid Waste collection sites and events. More than 98 percent of the population are provided paint collection service within 15 miles.
- PaintCare and local government programs sponsored 27 collection events including leftover paints. (This does not include Metro's HHW collection events.)
- More than 706,270 gallons of post-consumer paint was collected in 2020, despite clear Covid-19 impacts from March to June. Consequently, the 2020 collection volumes decreased by 8 percent versus 2019.
- About 6.7 million gallons of unwanted paints have been collected since program inception in mid-2010. This amounts to nearly 1.6 gallons of leftover paint collected per capita.
- Eighty percent of latex paint collected was turned into recycled paint plus another 3 percent of latex paints locally reused.

- Eight percent of oil-based paints were reused, and 92 percent was sent for energy recovery.
- Retail paint sales fees, which pay for the entire program, amounted to $5.6 million in 2020 with $40,000 sent to the state for oversight responsibilities.[6]

The Marketing Challenge

Marketing recycled content products comes with well-known challenges. There is frequently a perception that quality is compromised by having recycled content. Historically, some manufacturers might have offered products that really were shoddy, leaving a bad impression long after much better versions became available. This was the case with recycled content paper when it first appeared on the market decades ago, and certainly this has posed problems for those marketing recycled content paint.

Paint is particularly challenging because so much labor goes into applying it, and, simply put, people don't want to go through that labor again anytime soon! And, if anything goes wrong due to improper preparation, storage, or any other errors that can be made with paint of any kind, the recycled content is likely to take the blame. So marketing efforts must address quality concerns as well as many other considerations.

While a government agency does not typically have expertise in the area of marketing a retail product, MetroPaint staff, working closely with Metro's Public Affairs or communications staff, have stepped up to the challenge in a variety of ways over the years. On a few occasions marketing firms were hired to help out as well.

Target Customers

Some paint recycling advocates have focused marketing efforts on government agencies, believing that they should lead the way in proving the

6 Oregon Paint Stewardship Program 2020 Annual Report. 2021. Oregon Department of Environmental Quality. https://www.oregon.gov/deq/mm/Documents/pspAnnualReport2020.pdf (accessed August 22, 2022).

viability of the product to the larger market. This approach has not been successful for MetroPaint. There are significant barriers to selling recycled paint to the government. Even agencies with recycled content purchasing requirements have been reluctant to embrace recycled paint. Staff responsible for painting government facilities may be reluctant to try something new. Or those procuring painting services may have a hard time hiring contractors who are willing to use something they don't trust. And some government purchasers rely on rigid specifications that recycled paint may not be able to meet.

The great majority of the customers buying MetroPaint are do-it-yourselfers (DIY) painting their own homes, and to some extent smaller-sized painting contractors. Often they are repeat buyers and help to spread the word about its value. Considerable effort has been put into attracting larger institutional buyers, such as homebuilders and commercial builders, larger contractors, schools, public works agencies, but these efforts have met with limited success.

The customers for whom recycled paint is appealing tend to fall into two camps: those who are looking for a bargain and those who appreciate the environmental benefits of recycling. But in both cases there has to be a belief that the paint is going to do its job, protecting the painted surface and holding up for a while. And in the paint world, brand loyalty is a very big factor. Many painters, both DIY and professional, have chosen a brand that works for them and it takes an awful lot to get them to consider trying something new.

While the focus of MetroPaint marketing is targeting local and regional customers, a portion of sales are to exporters sending paint to overseas markets. More developed countries like the United States and the EU have advanced paint production technology, but much of the developing world does not, and the leftovers of paint collected in domestic collection programs, with appropriate quality controls, is of some value to them. However, exporters offer pretty low prices since they have substantial costs for shipping and distributing the paint, and customers who are not affluent. It makes sense that only batches that are surplus to domestic market capabilities, usually the less popular colors, go this route.

Pricing Considerations

As described earlier, there is a continuum of recycling processes and those lower on the sophistication scale tend to be lower quality and have a lower perceived value. On the low-end, some programs give their product away or maybe charge just a few dollars per gallon at a small fraction of the cost of new paint. The more quality control that goes into the manufacture, the more a recycled paint seller can charge. However, there is a limit to this, and even the best recycled content paint typically sells for significantly less than new. Thus, cost-conscious consumers are a key target market for recycled paint. There is also the notion that low cost equals low quality, so setting the price point for recycled paint products must balance these two considerations.

Competing with New Paint

MetroPaint, along with the recycled content products of some of the private sector paint recyclers around the country, is available for sale on store shelves, right next to virgin paint brands. Key marketing and sales challenges in this situation include

- color
- quality
- convenience
- environmental considerations

Color

Color is fundamental to paint. In the new paint world, customers typically have hundreds or even thousands of very specific shades to choose from. These days most paint is colored at the point of retail sale, adding custom tint to an unpigmented tint base for each purchase. Recycled paint incorporates all of the already tinted paint from collection programs, and out of necessity must focus on producing batches of a relatively small number of colors. So finicky customers may not find the specific shade they want.

One way to expand the color palette is to blend various colors of recycled paint in different proportions. While it is not practical for recycled paint retailers to stock a variety of ready-made blended colors, customers may want to try blending on their own. At the MetroPaint factory store a book showing samples of many actual blends is available for customers to peruse, along with supplies to facilitate them doing the blending at home. For projects needing larger quantities, preparing custom blends for them can be worthwhile.

Color consistency is an issue. As mentioned above, with smaller batches, reblending results in different shades from batch to batch. Larger batch sizes do result in more consistency, but people want to come back and get the exact same shade to blend in with the color they bought previously. With practice, batches can be made consistent by carefully blending in the right amount of the right colors of recycled paint to adjust each batch. This can be done by eye, and MetroPaint staff became very adept at this. The addition of sophisticated spectrophotometric equipment, while expensive, can greatly improve the color-matching process. Note also that consumers may have some unrealistic ideas about how well new paint color can be matched after the initial purchase, so some education on this topic can be useful.

Another big challenge is the mismatch between the volume of incoming colors, versus the demand for the recycled colors offered. White paint is very desirable, meaning that people don't tend to turn it into collection programs, so it is generally impossible to meet the demand. On the other hand people turn in lots of greens, gray, and browns but often don't want the limited selection of color shades that are available in the recycled paint. So disgruntled customers can't get the white they want, and large quantities of darker colors must be sold off very cheaply, or even disposed of.

Quality

Word of mouth is the number one way that potential customers are convinced of the quality of recycled paint and so sharing community member testimonials can be beneficial. Offering demonstration projects can also be useful.

Consumer acceptance is influenced by other elements that enhance the perception of quality. This includes packaging and labeling. High-quality, attractive cans and pails are optimum. For best results, labeling should be developed with the assistance of graphic design professionals and should include all of the label information needed, both legally required and helpful to the user. Warranty is another consideration. Paint consumers often look for long warranty periods on the products they purchase, a short or even no warranty can reinforce the perception of questionable quality.

While customers want to be confident in the quality of the paint they are buying, they seldom are interested in the technical details of performance characteristics. For this reason many will look to an external organization to provide an official certification. As part of the national paint dialogue described above, a certification approach targeted specifically to recycled paint was developed.[7] This involved both the Master Painters Institute (MPI), which is the leading paint specification authority in North America, and Green Seal, a nonprofit organization that certifies the environmental attributes of products. MetroPaint was the first product line that achieved certification under these standards after they were released. A small number of private sector recyclers followed later.

Note that some new paint products come labeled with other certifications, often with the word "green" in them, and many times, these are developed and certified directly by the manufacturer, that is, not an independent verifier. So a buyer beware approach is important for quality and environmental certifications.

For many years MetroPaint maintained its MPI and Green Seal certification, an endeavor that required considerable time, expertise, and financial investment. MetroPaint staff generally agree that this was very helpful in cementing the reputation of the product in the region but that eventually the various costs outweighed the benefit. Private sector recyclers tended to agree, and recently an association of paint recyclers has developed a more manageable paint quality certification system.

Convenience

An important element of successful marketing and sales is customer convenience. Initially MetroPaint was sold solely at the site of Metro's paint

[7] Green Seal, "GS-43: Green Seal Standard for Recycled Latex Paint." 2013. Green Seal. https://greenseal.org/wp-content/uploads/GS-43-Standard-Ed-1.2-04.2021.pdf (accessed August 22, 2022).

recycling facility in North Portland, the “factory store.” Since many paint consumers prefer not to travel far to get their paint, this was a significant limiter of sales. Over the years staff put considerable effort into finding resellers elsewhere. At first the focus was on resellers located elsewhere in Oregon and the Pacific Northwest, but eventually we wanted a better reseller network in the region. Early resellers included used building material retailers, environmentally preferable building material retailers, and a government surplus liquidator.

The big breakthrough came with the partnership with Miller Paint, as mentioned earlier. Between their own retail stores and other retailers that carried their product, including hardware stores and even a big box chain, MetroPaint's retail presence expanded considerably. This is an ongoing partnership and includes working together on things like retail display development and even our color production strategy.

One major limitation of the ability to recruit resellers was the limited supply of white paint. Resellers want to satisfy their customers, and many customers want white paint. Carefully rationing the white supply between the factory store and reseller locations only goes so far. At times we experimented with purchasing low-cost virgin white paint, from Miller and other sources, to stretch our supply of white paint, but this is an imperfect solution as it dilutes the environmental benefit of recycling and increases costs.

Environmental Considerations

Recycled paint presents a number of environmental benefits compared to using new paint, foremost among them a reduction in carbon emissions. Quantifying these impacts turns out to be very difficult. As part of the national paint dialogue, an in-depth life cycle assessment process was conducted. This turned out to be quite expensive, and the results were not conclusive, due primarily to the difficulties in the assumptions that are required to assess actual impacts across the lifecycle of both recycled paint and new paint.

When the public is thinking about the environmental impact of paint, however, they tend to focus on volatile organic compounds, or VOCs. This VOC focus turns out to be a complicated subject as well. To detail the complex considerations around these two environmental aspects, an in-depth fact sheet that was part of MetroPaint promotional material is included in the appendix. In reality many customers are not interested in delving into this level of detail, but it is good to have this document available for those who are.

Third-party certification also plays a role in assuring customers of the environmental preferability of the product they are purchasing, such as the Green Seal certification mentioned earlier.

And one more important detail in marketing any product with environmental benefits is understanding and complying with FTC (Federal Trade Commission) guidelines about the use of terms promoting environmental benefits.[8]

Marketing on a Limited Budget

Over the years, Metro's paint recycling program has tried out a variety of marketing avenues. Some worked well, others not so well. However, very little formal marketing research has been conducted, and so successful findings are anecdotal. Additionally, the marketing budget has always been limited, and certainly doesn't compare in magnitude or percentage of sales to the expenditures that new paint retailers make. The staff's belief is that word of mouth is the single most important element of convincing people to try recycled paint. When a neighbor has used it, and it looks great and holds up, and they brag about the money they saved, that can be pretty convincing.

Some of the many other approaches that have been utilized, always with an eye to maximizing results with minimum expenditure, are listed below. Some rely on a brochure containing basic information about the product and process, with representations of the standard colors available. They are listed roughly in order of our perception of how effective they may be.

- **Direct telephone and in-person contacts**. At times utilizing permanent staff, sometimes hiring temporary staff dedicated to marketing, and sometimes hiring contractors with varying marketing experience.

[8] "Green Guides." Federal Trade Commission. https://www.ftc.gov/news-events/topics/truth-advertising/green-guides (accessed August 22, 2022).

Generally focused on potential institutional buyers such as government agencies, contractors, and builders.

- **Marketing consultants**. These professionals can assist with logo and label development, maximize our partnership with resellers, develop overall strategies and campaigns, or create website content.
- **Website**. Create an online presence with a variety of content about the product, pricing, colors, and more.
- **Local partnerships**. Partnering with local governments and other collaborators in the region with similar goals, sending out information to their customers through newsletters, utility bills, and the like.
- **Earned media or publicity**. This is sometimes received in response to press releases, and includes television, radio, and newspapers.
- **Trade shows and presentations**. Booths utilizing a professional display can be effective at a variety of trade shows and conventions, particularly home-improvement or DIY-related, as well as at neighborhood fairs and other community events.
- **TV and radio spots**. These are paid for at the going rates for government agencies.
- **Demonstration projects and case studies**. Includes temporary signs to place in front of buildings being painted with MetroPaint, permanent plaques for placement in buildings that are open to the public where our paint has been used, and published case studies of projects utilizing MetroPaint.
- **Social media**. While MetroPaint does not have its own dedicated social media channels, the larger Metro agency social media presence has been used to disseminate MetroPaint-related messages, sometimes as part of agency-wide campaigns.
- **Newspaper advertising**. Used in the regional newspaper's "Home and Garden" section, as well as some smaller community papers, and industry and trade publications, sometimes including coupons.
- **Sales and coupons**. Although the paint price is very low to begin with, we have tried coupons and sales as a means of drawing attention from cost-conscious buyers.
- **Direct mail**. Mailings to a variety of government agencies, nonprofit organizations, low-income housing developers, painting contractors registered with the state, property managers, architect firms, and others.

- **Displays at permit offices.** Displays with brochures and fact sheet at building permit offices in nearby cities and counties can be effective.
- **Videos.** Videos of the production process and other details can be shown on cable access channels and online.
- **Internet sales.** We have offered MetroPaint through eBay.

The State of the Art Today

In 2005, MetroPaint moved into a 22,500-square-feet leased warehouse, and operations continue there today. While the central processing room is similar in design to the original 1999 custom-designed facility, over the years a wide variety of improvements have been implemented. In many respects the operation embodies the state of the art of paint recycling, although a small number of private sector paint recyclers have implemented more advanced technologies, such as robotic can opening and pouring, automated color scanning and sorting, and so on.

The ongoing quest for improving the MetroPaint production process continues, with recent improvements including more and larger tanks, industry-standard filtration, and routine use of colorants.

MetroPaint Facts and Statistics

Metro has been recycling latex paint since 1992, and as of June 2022 the program has

- opened about 15 million containers of leftover paint;
- processed 6.9 million gallons of paint;
- produced 4.3 million gallons of recycled paint;
- generated $19.1 million in revenue from recycled paint sales; and
- recycled 2.3 million pounds of empty steel paint cans and half a million pounds of plastic.

MetroPaint is available

- in twelve standard colors,
- in 1- and 5-gallon sizes as well as quarts,

- with a variety of specialty and limited edition colors available at the factory store,
- at a cost of $15 per gallon (MetroPaint factory store), and
- at more than fifty retail locations around the Pacific Northwest.

Costs and Revenue

Prior to the startup of the PaintCare system in Oregon in 2010, the paint that Metro collected from the region's residents was viewed in the same way as all of the other waste collected in the household hazardous waste (HHW) program, as part of a service that Metro was providing to residents, which came with costs. These costs were covered by a small percentage of the fees charged for solid waste disposal in the region, a common funding method for HHW management.

Careful analysis of costs and revenues determined that the net cost of each gallon of latex paint collected and processed for recycling was lower than any other disposal option available—as well as providing real environmental benefits. However, even with the revenue from paint sales it was not a money maker. Labor costs are a big factor, as each can of incoming paint is opened by hand and inspected; disposal of nonrecyclable paint is also significant. Equipment, supplies, and facility operating costs also figure in.

Over the years many interested parties from other regions contacted us, thinking that perhaps they could get into the paint recycling business, and we had to explain that the revenue from the sales of recycled paint is not sufficient to cover processing costs. Those private sector paint recyclers that were operating at the time were sustainable businesses only because HHW collection programs were paying them to recycle the paint that they brought in from their residents.

It was only when the PaintCare system began operating, and paying Metro as a contractor to recycle the state's paint, that operating costs were covered. For a number of years there was no net cost to the region's ratepayers. In other words, the combined revenue from selling MetroPaint and serving as a contractor to PaintCare covered all of the costs of operating the program. In recent years labor costs, marketing challenges, and other factors exacerbated by the Covid-19 pandemic, have resulted in some net costs to Metro once again.

The National Landscape

Since Oregon passed the paint stewardship bill in 2009 several other states have followed suit. As of June 2022, ten states plus the District of Columbia have paint stewardship laws, all passed with industry support, and all operated by the PaintCare organization. Several other states have made efforts to pass bills. Still, it does not look like there will be nationwide paint stewardship happening in the United States any time soon.

Where PaintCare is operating, there are convenient drop-off locations around the state, and HHW programs have a no-cost management option for the paint they collect. In all cases, recycling is the primary management method for latex paint, usually conducted by private sector recyclers not affiliated with the virgin paint industry, although there are a few municipally operated recycling programs in place, albeit on much smaller scale than Metro's program.

Another Case Study: A Successful Smaller-Scale Program at Tillamook County

For many local programs contemplating getting into paint recycling, developing a large, sophisticated process like Metro's is not realistic. An overview of a smaller, more streamlined program is included here as an example for HHW program operators who want to consider setting up something in their area. This is the program at Tillamook County, located on the Oregon coast, with a population of 25,250 people. The information is provided by program manager, David McCall, with operations depicted in Figures 2.2 to 2.5.

> We first tried out collecting paint for two months, and over the course of two days opened, mixed, and poured a lot of paint into a 55-gallon drum system. We then started our own permanent latex paint recycling program, aimed at providing a service to the community while reducing the quantity of paint that we'd have to ship off for treatment.
>
> Although we are able to send paint to PaintCare, our program's initial popularity enticed us to keep it going—for local benefit.
>
> We developed a local system, using materials we could procure locally. The only thing we had to bring in from elsewhere were the buckets and lids, as well as the blender. Intermediate bulk containers (IBCs) were used.

Each IBC is fitted with a funnel and two grates on the top. Paint is opened, mixed, and then bulked into seven different colors: white, off-white or ivory, blue, green, yellow, grey, and brown. (We don't keep red, pink, purple, and orange ourselves, but send them off to PaintCare, as well as oil-based paints.) When an IBC has 225–250 gallons in it we put the mixer on it and mix the paint in it. We run the mixed paint through another strainer or filter as it is poured through the bottom spout into buckets. The mixing and bucketing is done on off days, when no other activity is going on at the HHW facility. It usually takes about two hours to bucket out an IBC of mixed paint, and clean up afterward.

The system is flexible enough that we can make it mobile, though we have only conducted one mobile event since developing this system. As you can see from the numbers (see later), we have managed to keep about half of the paint we collect each year in our local latex reblending system. All our paint is used locally.

We have a program in place to provide nonprofits with paint at no cost, and we frequently offer a voucher for paint at local silent auctions. I would estimate that about half of the paint we sell each year is sold through our local Habitat ReStore, with the rest sold at a transfer station or our office.

For Habitat, this is a fundraiser as well. We only charge Habitat 12 dollars per bucket of paint. Habitat and our department charge the public 24 dollars per bucket of paint.

We were successful in selling or placing our annual quantity of paint until we topped 1,000 gallons. That happened to be a year when we had rain that lasted until about June, and no one was buying paint when we promoted it the hardest in April and May, but they were cleaning out their old quantities of paint! That created a backlog year that we still haven't fully recovered from, leaving us with paint on stock that we have to store.

Now that quantities have returned to more normal levels, I think that our balance will also return, and we'll better be able to market the annual quantities without having to store as much.

Our favorite colors tend to be gray and off-white (which we now refer to as "ivory"), though we tend to go through blue fairly quickly as well. The largest quantity we generate is brown, which is also the worst seller. (I guess everyone liked brown a couple decades ago, but no one wants it now!) One

Figures 2.2–2.3 Tillamook County paint recycling operations.

Source: David McCall.

of the main challenges is the inability to change the color, since we don't have the quantities like MetroPaint. But many people seem to prefer our paint and claim that it lasts longer than other latex paints they have used. And it seems as though the quantity that goes into a 275-gallon IBC results in a fairly similar average color on the output side.

Figure 2.4 Tillamook County paint recycling operations.

Source: David McCall.

We have honed our system into a fairly efficient system that works well with our HHW events program and provides a benefit that is well received in the community.[9]

[9] Personal communication with David McCall, Tillamook County, Oregon.

Figure 2.5 Tillamook County paint recycling operations.

Source: David McCall.

Some additional information about the Tillamook County latex paint recycling program: the paint is collected at two retail paint sites and one county HHW facility. At the HHW facility, about 1,500 gallons per year is reblended (the remainder is sent to PaintCare). Their paint is sold in 3.5-gallon pails for 24 dollars.

The Turn to Industry Responsibility

Thirty years ago, Metro saw the need for collection of leftover latex paint and the environmental and financial value of recycling as much of that leftover paint as possible. Twenty years ago the government and the paint industry in the United States entered into an in-depth dialogue process, ultimately resulting in industry agreeing to take responsibility, collecting the first can under their stewardship in 2010. The industry-run PaintCare program now covers ten states plus the District of Columbia, with work ongoing to adopt it in additional states. It will require many more years to pass legislation to cover the remaining states—or to see a federal law adopted—so local government collection programs will continue to shoulder the work in many areas.

Other approaches to managing leftover latex paint have been proposed, such as voluntary industry stewardship or advanced disposal fees collected by the state government to fund a state-run system, but the disadvantages of these approaches have caused them to fall by the wayside. This leaves just the two competing approaches: legislated producer responsibility or local government–funded collection programs (setting aside the significant portions of the country without any solutions in place).

The legislated producer responsibility approach provides a number of advantages. These include

- Sustainable financing,
- Convenient collection for consumers,
- Economies of scale,
- A level playing field for paint producers, and
- Opportunities for reducing toxicity and environmental impact.

There have been implementation challenges, many of them relating to the role of government oversight versus industry control of programs. But these challenges are being worked through and as a result, PaintCare is now operating more than 2,000 collection points and to date has collected more than 55 million gallons of paint. Interestingly, in many cases local government operations have been incorporated into the PaintCare system.

Summary: Key Lessons Learned

The overarching lesson conveyed here is that the government can lead the way. Starting in 1991, until the startup of PaintCare in Oregon in 2010, Metro was the only organization providing a solution to the leftover paint problem for the region. Further, Metro pioneered a recycling process that grew and innovated over the years, rivaling the best that the private sector can offer. Thirty years after first emptying cans of paint into drums and stirring them with a boat paddle, MetroPaint continues to be a vital part of the paint management infrastructure in the Pacific Northwest. Metro's successes helped bring about paint stewardship legislation in Oregon and elsewhere. Within the industry-led structure, smaller government entities that are willing to innovate can also help with solutions, as evidenced by the Tillamook County program.

As a government agency, Metro is quite willing to pass on what we have learned. Some key insights include

- Paint recycling is a continuum. Low-tech processes for consolidating latex paint can be launched at a very modest cost, but with increasing investment in equipment and staff expertise a higher-quality product can be made and sold, even one that can compete side by side with new paint on retail shelves.
- There is a net cost to recycling paint. Even when there is revenue from the sale of recycled products this is not enough to cover the costs of recycling. It requires either government-funded disposal dollars for collected paint, or ideally industry-sponsored funding under a stewardship system, to make ends meet.
- Using recycled paint in lieu of new paint does provide significant environmental benefits, among them being a real reduction in greenhouse gas emissions.
- Marketing recycled paint does present challenges, but it is possible to overcome them without a huge marketing budget. Attention to product quality, fostering positive word of mouth in the community, and utilizing third-party certification strategically are key approaches that can help achieve sustainable sales.
- When an industry steps up to take responsibility for its products, benefits accrue to multiple stakeholders. Producers better meet the needs of their customers. Consumers are provided more convenient access to properly manage products at end-of-life. And financial burdens are lightened for local and state governments.
- Industry and government can work together toward a common goal and the results can have a big positive impact on consumers and the environment.

A plucky government agency can certainly move the ball forward, but ultimately, for widespread convenient paint collection across the country and maximum environmental benefit, paint stewardship legislation is needed nationwide. Fortunately, the paint industry has concurred and is working with the government and other advocates to bring it about. Increasing industry support for the viability of paint recycling is one area where attention should be focused. The value and viability of paint recycling—as powerfully illustrated by the successes of Metro's program—provide a strong argument for why more states should adopt paint stewardship legislation.

3

In the Trenches: Getting EPR Done

Heidi Sanborn and Tim Goncharoff
National Stewardship Action Council

"Waste" Management: How We Got Where We Are and Why We Need to Change

For most of our history, people didn't give much thought to the messes we made. When things got cluttered, or resources ran short, we'd just move on to a new place. People generated "waste" that was largely natural and biodegradable. Food remains and other organic materials were fed to the chickens, pigs, and goats. Other product waste, such as paper and remains from leather goods, needed disposal. There was usually a handy river into which refuse could be dumped or space somewhere on the outskirts of the settlement where holes could be dug to pile in the waste to decompose. If full, these holes often became "burn dumps" to make more room for discards. There were no materials that nature could not "digest."

And yet, waste management has always been a major issue with which people have had to contend. As we began settling in bigger towns and cities, the problem became more acute. Before the automobile was invented, horsedrawn carriages were the source of manure that littered the streets of major cities. And while open refuse dumps were commonplace in the 1900s, they came with problems, such as vectors, groundwater contamination, and rainwater-penetrated leachate that coursed into nearby bodies of water. While problems remained, we got smarter with time. In the late 1930s, the first modern sanitary landfill was built in California, followed by increasing and eventually widespread use of this practice in the United States in the 1960s and 1970s. Still, we've always been keen to move waste as far away from our communities as possible, whether it be to the outskirts of town, to other municipalities, or to other countries.

With a shift from largely agrarian and handicraft societies to a modern society transformed by technological changes and innovations and industries dominated by manufacturing during the Industrial Revolution, things changed. People now had new processes, chemicals, and materials unlike those from nature. This included plastic production that dramatically grew during the Second World War. Many other products, such as the pesticide DDT, were developed during wartime and produced postwar for wider public use. Neither safe for the environment nor safe for human health, we continued to churn out a wide array of harmful products for distribution on the consumer market for years to come, with such products as mercury thermometers, lead gasoline, and more.

As the human population grew exponentially, thanks to the Industrial Revolution, so did the production of goods. Many products were now sourced from nonorganic materials and combined into "horrible hybrids": products that contained different materials and were not designed for separation and reuse or recycling. Waste products were also more challenging and expensive to manage due to their toxicity.

In 1991, the US Environmental Protection Agency (EPA) passed Subtitle D rules (OSWFR-91-004) that became effective in October 1993.[1] This established municipal solid waste landfill (MSWLF) design criteria that included siting restrictions, performance standards, operating practices (e.g., daily cover), groundwater monitoring, and other protections. Local governments started building the newly engineered sanitary landfills, as well as wastewater treatment plants to process leachate.

1 Federal Register: October 9, 1991, Part 2. 40 CFR Parts 257 and 258. Solid Waste Disposal Facility Criteria; Final Rule

A Change of View

The world eventually began to see Earth a little differently when, in 1959, the first photo of Earth from space was taken from Explorer 6. In 1968, a photo of our small planet taken from the moon by astronauts aboard Apollo 8 further affected our perspective on Earth: it became explicitly apparent that we live in a closed, isolated system. It became more apparent that we needed to care for our world. Waste products are part of the system and an aspect that we need to manage as best as possible. In fact, there is no "away": the idea of throwing trash "away" was no longer tenable.

And then there's the problem of throwing trash away only to have it return! Consider the garbage barge *Mobro 4000*. In 1987, the vessel departed Islip, New York, carrying more than 3,000 tons of garbage to dispose of its cargo elsewhere, without success. After over six months of national and international port rejections, *Mobro 4000* returned to the New York City area to have the waste offloaded from the barge and incinerated. This highly publicized debacle underscored the problematic nature of municipal solid waste management.

Engaged in the collection and management of municipal solid waste (MSW) are myriad federal, state, and local governmental agencies, nongovernmental organizations, businesses, individuals, and others. We are all familiar with the well-orchestrated curbside pickup of our trash, or drop-off centers to which we haul our never-ending waste materials and products. Residents are taxed and pay fees for waste management services.

In the 2000s, the United States began to look further upstream—toward producers—rather than downstream, at the municipalities and individuals who bore the burden of unwanted and leftover goods at product end-of-life. Producers, after all, create many of the consumer products that are difficult to "de-manufacture" for recycling, contain multiple materials increasing the complexity of material recycling, and incorporate toxics into their formulations and designs. We entered this realm of consideration higher on the learning curve, thanks to the efforts of our European colleagues who in the 1980s had already begun to change their policy direction toward the role of producers through product stewardship and Extended Producer Responsibility (EPR).

Product stewardship is different from EPR. Product stewardship comes in many forms, examples include visible fees managed by the government or a producer voluntarily reducing the impact of their product, such as eliminating a toxic chemical from the product to make it easier, safer, and less expensive to manage or recycle.

An example of product stewardship is California's government-run tire program. Since 1990, Californians have paid an upfront disposal fee with each tire purchase. The state agency known as CalRecycle, formerly the California Integrated Waste Management Board (CIWMB) and subsequently the California Department of Resources and Recycling (CalRecycle), is mandated to regulate and manage waste tires, and they use those funds to collect and properly dispose of the tires.[2]

California, like many states, has had its share of problems with outdoor tire stockpiles. Improper storage such as this can have catastrophic results. In September 1998, "the largest tire pile in the West" caught fire by lightning strike and seven million tires burned for a year spewing toxic smoke into the air and resulting in ash that settled on crops and in waterways in California's agriculturally productive central valley.[3]

While more widespread effort is needed to reuse the materials that comprise tires, some options are better than others. Whereas rubberized asphalt concrete is a good use, safety concerns arise when rubber tire chips for playground surfacing are used. Although product reuse and recycling are part of the product stewardship picture, a better option is EPR, which requires that producers internalize the cost of managing their product at end-of-life and take responsibility for the system and its impacts—including the safety of the recycled products.

Taking a Lead

Two decades ago, the California Product Stewardship Council (CPSC) was established to better deal with waste and pollution. Instead of placing the burden

[2] Tire Management Overview. *CalRecycle*, 2021. https://calrecycle.ca.gov/tires/overview (accessed 2023).

[3] Walt Wiley, "Calif. Sues over Last Fall's Tire Fire." *Tire Business*, April 11, 2000.

on taxpayers and local governments to clean up after polluters, the council supports holding the producers themselves responsible. After all, if producers don't bear the costs of waste management, cleanup, and disposal, what incentive do they have to be less wasteful and toxic producers? The time was right for CPSC but the process was far from easy. The once foreign idea that producers might be held accountable for products at end-of-life was beginning to take hold. After a number of successes in California, requests came in for help from far and wide. Eventually a new organization was formed—the National Stewardship Action Council (NSAC)—to tackle the issue and take the battle nationwide.

Let's consider light bulbs. It's clear now, for example, that fluorescent bulbs were a transitional technology, and that dependence on a product that is made of glass and that contains mercury was never going to be a smart choice. But in the early 2000s, everyone was using incandescent bulbs and wasting large amounts of energy. While light emitting diodes (LEDs) were a dim glow on the horizon, fluorescent bulbs were an energy-saving technology that was readily available on the market, and many agencies and environmental groups went all-in to support the transition, attracted by the idea of saving money and energy. The problem is that if you want to greatly increase the use of a product that is toxic (fluorescents contain elemental mercury, released as a vapor when glass bulbs are broken), you'd better have a plan for proper waste disposal. After all, "green" technologies aren't actually green if they have a big toxic waste disposal problem at product end-of-life.

As often happens, numerous states banned fluorescent bulbs from disposal in landfills, including California in 2006, but offered no funding to manage them, leaving local solid waste operations to pay for it on their own. The cost to properly collect a bulb and sequester the mercury was at the time about equal to the price of the bulb purchased new. This made it impossible for a landfill ban to be enforced because no one could afford proper management. A friend coined this as a "ban without a plan" which caught on in California, making EPR seem like the logical policy tool. And with consumer take-back, returning it to the purchase location is the most intuitive and convenient for consumers. Bring your burned-out bulbs back to the store when you buy new ones, and they'll be properly disposed of, at no cost to you. A good idea, but not perfect.

Some products, such as fluorescent bulbs, single-use vaping devices, and one-pound propane gas cylinders, are so toxic and wasteful that they should just be banned from sale—especially when there is a less hazardous and more sustainable alternative.

In support of a circular economy, the first tenet of which is to reduce waste and pollution, the NSAC introduced, passed, and signed into law AB 2208 (Kalra) in 2022 to ban fluorescent light sales in California by 2025. Fluorescents will be replaced by more efficient LEDs that do not contain any mercury.

As we moved away from early legislative efforts that focused mostly on the disposal end of the product cycle, our approach broadened to consider a product's life cycle more fully—everything from resource extraction to manufacturing, distribution, sales, use, and recycling or disposal. If we really want to make change, we need to deeply understand these issues. We have learned a lot of policy lessons along the way, starting in 2008 when we passed the first EPR bill in California on mercury thermostats.

In 2003, I [Heidi Sanborn] worked as an independent consultant and in contract with several local governments to draft reports on how they could drive EPR from the local level up. I was also the lead consultant with the Product Stewardship Institute (PSI) to facilitate the national paint dialogue; this PSI project led to the completion of eleven projects that in turn informed the development of the model paint stewardship legislation. This experience informed my understanding of industry groups and government and in significantly designing a product stewardship program.

Later, I had the opportunity to consult with the California Integrated Waste Management Board (CIWMB) on the first report on EPR in California, titled "Framework for Evaluating End-Of-Life Product Management Systems in California," in June 2007.[4] A significant result was that CIWMB adopted an "Overall Framework for an Extended Producer Responsibility System in California"[5] in 2008 after taking input from local governments and many others during multiple public hearings.

[4] "Framework for Evaluating End-of-Life Product Management Systems in California." CalRecycle, June 2007. www.calrecycle.ca.gov/Publications/ 1-800-RECYCLE (California only) or (916) 341-6300 Publication # DRRR-2010-013.

[5] "Overall Framework for an Extended Producer Responsibility System in California." CalRecycle, 2008. https://www2.calrecycle.ca.gov/Docs/Web/112569.

In 2007, I was hired as grant writer to obtain a household hazardous waste (HHW) grant from CIWMB to fund an EPR education initiative for local elected officials and staff. At the same time, I worked with others to form the California Product Stewardship Council (CPSC) and served as the director alongside Carol Misseldine for a couple of years. Through CPSC, we worked with local governments to increase education on product stewardship and EPR and ways to support these policies locally and at the state level. Rob D'Arcy, who was the Santa Clara County's HHW Director at the time, with the support, and at the request, of many California local government leaders, facilitated planning meetings to form CPSC as an official 501(c)(3) organization tasked specifically with EPR education.

The simple mission was to turn our approach to waste and pollution into becoming more of an upstream approach focusing on producer responsibility. After all, if producers do not bear the costs of their product disposal at end-of-life, what incentive do they have to make products less resource-intensive, toxic, and waste-generating? The time was right, and the approach had been demonstrated to be effective, for many years, in Europe, Canada, and other countries. Under the direction of CPSC's Executive Director, Doug Kobold, the mission remains much the same, "supporting policies and projects where producers share in the responsibility for managing problem products at their end of life."[6]

In 2015, the CPSC board and others formed the National Stewardship Action Council (NSAC) as a 501(c)4 nonprofit to take the vision and, as needed, the fight for EPR nationwide. Our goal? Less waste, less pollution, smarter production processes, and a greener and more equitable planet for us all. NSAC includes government, nongovernment, and business constituents as well as individuals who advocate for shared responsibility, in striving to achieve an equitable circular economy. Using the definition of circular economy from the Ellen MacArthur Foundation,[7] this is a three-part test to (1) reduce waste and pollution at the source, (2) keep materials in motion, and (3) regenerate natural systems.

[6] California Product Stewardship Council. https://www.calpsc.org/about-us.

[7] "Circular Economy Introduction." Ellen MacArthur Foundation. https://ellenmacarthurfoundation.org/topics/circular-economy-introduction/overview.

In this chapter, we provide a how-to approach based on our experience, to develop a potentially winning strategy to draft, introduce, pass, and implement EPR legislation, including key decisions that should be made and in what order, to "get EPR done." The caveat is that there are differences and nuances depending on the products, producers, states, and a multitude of other variables, and so you should proceed with caution. Recognize that what works for some may not work for all and the recommendations provided here offer a starting point from which to deviate and adapt to suit your own needs.

Know Your Playing Field and Be Prepared for Game Day

It is tremendously helpful to begin with a few key questions. These will help you to identify if in fact there is a problem and possibly the extent of it, establish your issue priorities, determine if there is a need for legislative action and when the timing might be right, and develop a strategy. Let's begin.

Can You Substantiate the Problem?

We all know folks who insist that everything is fine. "All products are produced in a safe and healthy manner." "The government wouldn't allow a dangerous product on the market." "Pollution concerns are overblown." "Consumers benefit from the products and don't mind having to pay for disposal." "No problem, technology can solve anything."

Maybe they genuinely believe there is no problem. In issue discussions, strive to meet people where they are. In other words, base your argument on elements of their knowledge and awareness so that they have a foundation on which to build. Once you can demonstrate that there is indeed a problem—one that's coupled with a solution, such as where producers have an important role to play in ameliorating issues associated with their products—you can share information on EPR and its potential as a solution to problems associated with a product.

Take for example, again, the issue of fluorescent light bulbs which contain mercury. We first gathered data on the effectiveness of household hazardous waste (HHW) collections, a primary location for collection of consumer products with hazard properties, including fluorescent bulb take-back. To do this we began with an informal look into convenience factors that might make it challenging for residents to attend collections. We determined that in our area of California, it generally wasn't easy to find such HHW collection information as facility location, hours, and if reservations are needed to bring items. Some HHW collection facilities are inconveniently located in terms of proximity, thus necessitating extra gas expenditure. In terms of hours of operation, some conflicted with work schedules, hours required more of a time demand for residents due to long on-site vehicle queue lines. There is no commonly agreed-upon method to measure how much HHW there is and how much we collect, but there is a common belief in the industry that there is a lot of HHW that is disposed of improperly in the trash. We think this limited reach is a shortcoming in the process. It also implies that alternative, less desirable behaviors are occurring such as disposing of HHW in the trash, on the ground, down the drain, or into storm drains.

We would present this data to the legislators, along with facts. One example was that Sacramento County allocated $50,000 in one year to collect mercury-containing fluorescent lamps (bulbs). They reached that expenditure limit early and had to stop promoting that they had accepted them, given lack of funds to maintain collection. They couldn't choose to raise rates high enough to collect more lamps, even from residents who were willing to "go the extra mile" and drive to an HHW facility. These combined factors made a very compelling argument that a problem existed that legislators should recognize. But the problem needs further substantiation, and visual confirmation can be powerful.

We documented that many fluorescent bulbs ended up in the trash and landfills about the time, which could put materials management industry workers at risk of exposure to mercury. Photo evidence showing the bulbs in a landfill taken after the mercury lamp "disposal ban" was implemented in California was compelling to legislators.

Another data point we could offer was the demonstration to retailers that it was worth their effort to collect lamps from customers for recycling. Retailers were reluctant to get involved for fear of issues associated with accepting a used, toxic product in their stores. To convince them to take them back, we reminded them that brand new fluorescent bulbs on their shelves also contained mercury and that providing a safe disposal method benefited their customers. Further, product take-back could increase customer foot traffic and store loyalty, that is, a value-added service. We showed them studies that made the point. Providing a recycling service such as this would bring people into the store who would hopefully buy more bulbs and maybe other products while there. After the data helped to convince one store to accept returned lamps, more stores jumped on the bandwagon and those without this service were likely at a competitive disadvantage. This depended on the existing availability of service for the store's own lamps, and the cost of the mercury lamp recycling service, which can be expensive.

Have You Clearly Established Your Priorities?

We all have priorities, and these priorities shift for various reasons due to things such as crisis management. When you delve into the world of the circular economy, a systems approach to sustainable materials acquisition, use, and materials management, you quickly learn that this comprises several priorities along different, connected pathways in the system. You could, for example, prioritize working on toxics source reduction by promoting collection, or banning the sale, of mercury-containing light bulbs as demonstrated; promote recycling of a common household product such as batteries (e.g., mercuric oxide or lithium ion); or strive for packaging EPR that includes several items, such as source reduction and toxics reduction, refill, recycling, and recycled content. This is what we did in passing SB 54 (Allen) in 2022, when we rolled everything that we learned from the bills that passed in California and across the nation into one packaging EPR bill.

We engage in this circular economy (rather than the linear approach of "take, make, waste") because we strive to effectively reduce resource consumption and waste generation and simultaneously protect health and the environment. We recognize that there is no waste in nature because everything is feedstock for the next process. While humans have created waste as we know it, we want to change the paradigm from "waste" to "materials" that flow back in the circular system. Maximizing positive impacts and minimizing or eliminating negative impacts are the goals. Still, human behaviors are inconsistent and contradictory with these goals and there are competing demands for our attention. For instance, if state lawmakers are overwhelmed with a massive budget deficit, they may not be receptive to a new and complicated producer-run program. It's all about timing and legislators' energy or capacity to deal with the situation. That is, their bandwidth.

Politics always come into play. If local officials are leery of offending the business community, especially local businesses, and chamber of commerce groups, they are less likely to embrace anything that seems like a "radical" new approach. And most working people are chronically short of resources, working far too many hours on too many projects and they are not able to tackle something new and complicated, let alone controversial. Therefore, we need to make smart choices on what to ask legislators to take on.

Ask yourself, "What are the priorities?" and "What can I get done where I am, with the resources I have available, in this political situation?" This exercise is not a license to aim low, rather it is a call to be bold and innovative. We won't change the world by doing things the same way we've always done them. Changemaking necessitates thoughtful and strategic goal prioritization. We must consider what is both important and urgent, and the impact and consequences of action versus inaction. Pharmaceutical drug management provides a good case in point, with valuable lessons on how to pass difficult EPR laws.

Pharmaceutical EPR: They Said We Couldn't Do It, but We Did!

Americans, with just over 4 percent of the world's population, take almost 45 percent of the world's pharmaceutical drugs.[8] The average American

[8] "Biggest Pharmaceutical Markets in the World by Country." *World Atlas*, 2017. https://www.worldatlas.com/articles/countries-with-the-biggest-global-pharmaceutical-markets-in-the-world.html (accessed 2023).

takes four prescription medications[9] plus countless over-the-counter drugs, supplements, and the like. Our medicine cabinets overflow with pills, many of which are often leftover and expired.

Prescription drug abundance, stockpiling, and accessibility have led to serious problems: accidental ingestion by unintended users, overdoses, drug misuse and abuse, and black-market sales. Even conscientious efforts at disposal of leftover, unwanted prescription drugs are problematic. Should consumers flush unwanted medications down the drain or toilet? Throw them in the trash? Or, before putting them into the trash, mix them into used cat litter so they're less desirable to trash pickers. Clear answers have been elusive to consumers and the impacts of misguided decisions impact people and the environment. Fish and other aquatic organisms absorb drugs from wastewater, as demonstrated in a Florida study that found an average of seven prescription drugs in the fish tested, and as many as sixteen different drugs in a single fish.[10] Aquatic animal behavior and physiology are impacted by our excreted and flushed drugs.

These issues outline the problem statement we used to drive home the need for California to adopt EPR for medications. It was obvious to us, especially after being a part of a panel with former senator Wesley Chesbro and CIWMB chair Margo Reid-Brown at a biotech industry meeting. The pharmaceutical industry would have to play a significant role in the solution, although they did not yet acknowledge the problem. And so the industry was offered an opportunity to partner on pilot projects to address medicine and needle disposal issues. Only one company out of about seventy in attendance showed interest in that work, and later that one company called to say their attorneys advised them against participating. We think it was because even working with us on pilots would imply that they had some responsibility to tackle their pharmaceutical drug and needle waste.

9 "Too Many Meds? America's Love Affair with Prescription Medication." *Consumer Reports*, 2017. www.consumerreports.org/prescription-drugs/too-many-meds-americas-love-affair-with-prescription-medication (accessed 2023).

10 Teresa Hornstein, "Study Reveals Fish on Drugs in South Florida Waters." CBS12, May 19, 2022. https://cbs12.com/newsletter-daily/study-reveals-fish-on-drugs-in-south-florida-waters (accessed June 13, 2025).

With the pharmaceutical and needle manufacturing industries clearly denying the problem, few elected officials had interest in "taking them on" to require that they share responsibility for products at end-of-life.

Some local governments then tried meeting with representatives of the pharmaceutical companies to ask for help, and those who tried were initially met with industry stonewalling. It was time to try a *local* approach.

A few California cities and counties decided to take the leap and take on Big Pharma. They worked together and developed and passed true EPR ordinances for pharmaceuticals. The first county, Alameda County, was the bravest. That's where the City of Oakland is located and where they have had a prescription drug abuse epidemic. Alameda is on San Francisco Bay and researchers determined that wastewater effluent tainted with medications was impacting the bay ecology, so they had many reasons to implement an EPR ordinance for pharmaceuticals. Under the new local ordinances, first passed by Alameda County in July 2012 and soon followed by San Francisco and twelve other local governments, the pharmaceutical companies would have to create and fund a collection program for unused medicines and properly dispose of them at no extra cost to the consumer. Of course, there is much more detail and nuance to these laws than is indicated here. To understand how we won and got them passed, it's important to know the arguments on both sides.

To demonstrate the back-and-forth of Alameda and supporters versus Pharmaceutical Research and Manufacturing Association (PhRMA), here are some quotes from press in 2012:[11]

1. "The county should not be responsible for continuing to bear the financial burden alone," said Nate Miley, president of the Alameda County Board of Supervisors and sponsor of the ordinance. "It could be a model ordinance that everyone could use, not just in this state, but throughout the country."
2. "If they can create drugs that save our lives, I'm confident they can figure out ways to get pills back," said Heidi Sanborn, the executive director of the California Product Stewardship Council.

[11] Lisa Aliferis, "Alameda County Takes On Drug Companies over Medication Disposal." *KQED*, July 16, 2012.

3. "Wastewater plants are major pathways of pharmaceuticals into the environment," said Melody LaBella, pollution prevention program coordinator at the Central Contra Costa Sanitary District. "Much of that pollution comes through humans excreting the drugs after ingesting them. Some of it comes from unused medications that are flushed down the toilet. Wastewater treatment plants are not equipped to remove the drugs. Unwanted pharmaceuticals are the proverbial low-hanging fruit."
4. "There is no evidence that this will affect the number of poisonings in the community," said Ritchard Engelhardt, vice president of government affairs for BayBio, a trade association representing the life science industry in Northern California.
5. Consuelo Hernandez, vice president of government affairs for the California Healthcare Institute, a trade association that represents 300 biomedical companies, universities, and research institutes, said in an interview that her organization "doesn't have any plans to sue," but she added: "We can't predict what we will do in the future."

Unsurprisingly, the pharmaceutical industry sued in the groundbreaking case *PhRMA v. Alameda County*. Alameda County prevailed in the first trial, upon which the industry appealed to the Ninth Circuit Court of Appeal, alleging, among other things, that the actions of Alameda County and other jurisdictions were an unconstitutional interference in interstate commerce. The transcripts make for fascinating reading.[12] It was during this court case that 1,500 of our emails were subpoenaed; they found nothing. The important takeaway is that Alameda County won again and helped lay the foundation and "rules of the legal road" for counties around the nation to establish EPR for product waste.

The appeals court said a few very important things.

In response to PhRMA's argument that local governments had no right to regulate national or even global companies, the court said (paraphrased): Of course they do. Ensuring the health and safety of their residents is among the most fundamental responsibilities—priorities—of local government. Faced

[12] "Pharmaceutical Research and Manufacturers of America v. County of Alameda." *FindLaw*, September 30, 2014. https://caselaw.findlaw.com/court/us-9th-circuit/1679319.html.

with the impacts of dangerous products, they are well within their rights to regulate them. The industry then argued that if Alameda County was allowed to make this infringement, other jurisdictions would likely follow. Again, the court responded along the lines of (paraphrased), Well, why wouldn't they? Lastly, the manufacturers argued that if local governments were allowed to regulate pharmaceuticals, other hazardous products were likely to follow. The court's answer was more or less, Sure, why not? And of course, local governments everywhere took note. PhRMA appealed to the US Supreme Court, but in 2015, the court declined to hear the case, meaning that the decision of the appellate court is now the law of the land. EPR is legal so long as the product has a health and safety nexus.

Emboldened by the decision, numerous other cities and counties rapidly passed similar laws, each of them a little different to meet other recognized needs. Some included pet meds, others added needles ("sharps"). Some required a certain number of pharmacy drug take-back locations, while others said all pharmacies must participate. Finally overwhelmed, the industry came to the negotiating table seeking uniformity, helping to craft a new statewide law that is now rolling out through California. Other states have taken similar action, although California remains the only state to pass EPR for sharps, and more are sure to follow. We would encourage all states to tackle both sharps and medications under EPR at the same time. This is not just because both products are part of a national public health problem, but because many of the same companies, retailers, and regulators address both product types.

The takeaway: The problem was clearly and definitively identified. It was a high-priority product, in urgent need of action. The consequences of inaction were too great and were not just one area, like a "waste" problem. It was a problem for waste, water, public health, worker safety, and the environment. This aligned with the interests and efforts of many different stakeholders willing to fight for these ordinances, including seniors, waste management unions, waste haulers, clean water groups, some UC Berkeley students who saw friends taking illegally obtained Adderall (such as from medicine cabinets), and so many more. The responsible parties were brought to the table with key stakeholders and solutions were found and implemented—on a grand scale.

Is the Problem Ripe for a Legislative Solution? Is the Timing Right? Might There Be Unintended Consequences or "Blowback"?

How many environmental organizations can you name? Probably dozens, or maybe a hundred. There are a lot of them, with lots of different angles, overlaps, big goals, and effective action. They serve a vital role in protecting our (and other species') health, safety, and environment. It's the farthest thing from a monolithic group that agrees on everything. The same goes for companies and their associations. Our advice is that you consider working with companies and environmental groups separate from their associations. It's hard to say that "business opposes the bill" when the business community is not a homogeneous lot. Find individual leaders (in business and NGOs) and win them over to your team. Remember that EPR is difficult to pass, and your strategy should always include support from one or more key members from each stakeholder group. This is because when you bring a legislative proposal to an elected official, one of their first questions will be, "Who's for it and who's against it?" Make sure that you have a leader who is a politically influential player, from each of the big stakeholder groups for that product type, on your supporters list.

Decision makers also need to understand the gravity of the issue—they need data to make the case for the problem statement. To make progress on a priority issue, one way to make progress is to ask stakeholders for stories. Encourage people to tell their stories first-hand at hearings and meetings, to make the problem real and if emotionally impactful, not to be concerned about pulling at the heartstrings of decision makers; they will more likely remember the story and it will change votes. Qualify and quantify the problem. Data and pictures speak volumes. The following strategies can also maximize your chances of success in garnering issue attention:

- Prepare lobbying materials: use factsheets and letters to explain why legislation is the best solution.
- Find allies: shop for bill authors and co-sponsors, line up endorsements, schedule press events, and even hire a press consultant if you can afford one. Come out of the gate strong, supported, and ready to go.

- Network: tap into your partners and their networks and build grassroots understanding of the problem statement and support for the solution.
- Find a professional lobbyist: if possible, use a good professional lobbyist with access to the highest levels of government.

The key, based on exhaustive homework and support-building, is knowing if in fact, the problem is ripe for a legislative solution.

Solar Panels

An example that brings this home is the solar panel industry. Solar panel (photovoltaic, or PV) installations have increased in the United States since the 1970s. They show up on rooftops and in fields and parking lots, and everywhere from rural barns to the White House in Washington, DC. With concerns over the climate crisis and high energy costs and with the costs (including incentives and rebates) falling for solar panels, the adoption rate for panels has increased—they're being installed more now than ever. Photovoltaics have two primary designs that can include metals, such as copper, lead, cadmium, and aluminum, and plastic, among other materials. But since solar panels only last twenty-five to thirty years, what do we do with old panels? The first generation of PV panels is being retired, and in California, there are regulations for disposal at product end-of-life. Panels are big and heavy, and they don't come in ones and twos—they arrive at landfills by the truckload. Despite increasing numbers of PV panels now at product end-of-life, recycling opportunities are scarce and expensive. They may need to be shipped hundreds or thousands of miles, at exorbitant rates, for recycling or, worse, for landfill disposal.

Solar photovoltaic panel EPR serves as an answer to this problem. With valuable solar panel components, producers benefit from collecting and recycling them. This again presents a win-win-win situation by (1) shifting the burden of materials management from residents and local governments, (2) affording producers an opportunity to reuse materials and create customer loyalty, and (3) solving environmental problems. EPR for solar panels has been enacted in Europe for many years, but in the United States only the State of

Washington (2017) has passed and implemented a program, and the County of Niagara, New York, passed an ordinance in June of 2021. The precedent is set, thanks to heavy lifting by environmental groups, local and state governments, and eventually, the industry itself.

Beverage Container Deposit Laws (Bottle Bills)

Consider another example: returning beverage containers such as glass and plastic bottles and aluminum cans. This is legislation that requires an upfront deposit fee (e.g., 10 cents on the purchase) that is refunded when the consumer returns the beverage container to a store that sells the product, a redemption center, or "reverse vending machine." The benefits of a system like this are apparent: it's a good way to recapture material for recycling, it reduces waste disposal, and it decreases litter (due to a financial incentive to return bottles). Producers (i.e., beverage manufacturers) share the responsibility for their products at end-of-life through this practice. (For more information, visit https://www.bottlebill.org/index.php.)

Is the timing right now for beverage container EPR at the national level? A national bottle bill would be the first EPR bill at this level. We think it is the right time because there are ten states that already have bottle bills, some for over two decades, and *a lot* of valuable materials could be spared from waste disposal by having bottle bills in the other forty states. Bottle bills are generally producer- and EPR-run. They are "proven policy" and we have learned lessons from the history of bottle bills. Large corporate leaders, like the Coca-Cola Company, have made pledges to use recycled content and rely on bottle bills for material capture. They may not be able to meet their recycled content pledge if they cannot get the material back from the public in the next couple of years. Many large companies and associations are now calling for a national bottle bill, with comments such as this from Tom Dobbins, President and CEO of the Aluminum Association, Washington:

> Container deposit programs are a key element of a strong aluminum scrap supply chain. A recent analysis by environmental research firm Circular

> Matters showed that while the deposit states consume about a quarter of all beverage cans, they generate more than a third of all cans recycled. It is important that we find a way to keep this vital material flowing while maintaining strong public health guidelines.[13]

If your state does not yet have a bottle bill, this may be the best EPR bill to introduce first, despite challenges in passing any EPR bills. This is because there is a lot of corporate support and supporting data and model bills. Also, because it would help further develop the collection and sorting infrastructure that can sometimes be used for other materials.

Have You Carefully Designed Your Strategy?

To affect change, we need to understand the legislative process and develop strategies to reach intended goals. Most importantly, we need to have a keen sense of human interest, intent, and behavior. Only then, as we begin to understand stakeholders' intrinsic and extrinsic motivations, can we build an effective strategy and momentum for enduring change.

Before deciding it's time to introduce a state bill, a key question to ask, especially given this is a big commitment, is, *can it pass*? If you know the odds are against the bill passing, you might want to reconsider and alter course. A lesson we learned is that it is sometimes better to push for local ordinances, as we did with pharmaceuticals, and keep working at passing them until the industry agrees to negotiate at the state level. Particularly when public health and safety are on the line, you will possibly have luck in passing local ordinances. The local elected officials are much closer to their constituents than those at the state or federal levels. Also, the reality is that large associations often have less connection, and thus perhaps less credibility, with locally elected officials, than they do with federal or state officials. Further, local officials may see associations as obstructionist (that also rarely give to local campaigns). All this likely makes for a much easier first-time passing of local ordinances, especially

[13] Deanne Toto, "Trade Associations Voice Support for Bottle Bills." *Recycling Today*, April 20, 2020.

if you are up against a well-funded and oppositional national industry group, as in our case with Big Pharma.

In terms of working to pass a state bill, even when you have determined the time is right to introduce it, be very careful—especially if you feel that the chances of the state bill passing are very low. Also take care regarding the amount of pressure you exert. If you push too hard and too fast, and the bill is voted down, that could spell trouble. The legislators who vote against it are "stuck" with that vote and probably don't want to be labeled as "flip-floppers," so they won't change position in the future. It may be best to hold public hearings where the producers are on record stating that they understand the problem and respond when you ask them what they plan to do about it. Public campaigns that connect their brand to waste or pollution, also get their attention.

It behooves us to consider what it takes for a bill to become a law, as shown in Figure 3.1. It is a long and complicated process, involving many steps. In most states, thousands of bills are introduced each year. Most will not get past the first committee, fewer will pass both houses, and even fewer will be signed by the state's governor and enacted into law.

Know Your Policymakers, Before You Ask Them to Carry Your Bill

Making lasting change means delving into the political realm. Find out where your elected officials stand on public health, social, and environmental issues. As with other things, it's important to do your research so that you know the facts, who you're dealing with and their priorities. What issues did they champion in their campaign? What does their website have on it? Who are their staff and what are their positions on topics? Who are their main constituents, allies, and funders?

Which way does your local community generally lean in terms of support for environmental protection? Is the dominant view that environmental protection is a cost to businesses? This view (of business protection) is not unusual; frequently there is suspicion that environmental regulation is hostile or costly to business and arises mainly from the far-left caucus. Consider

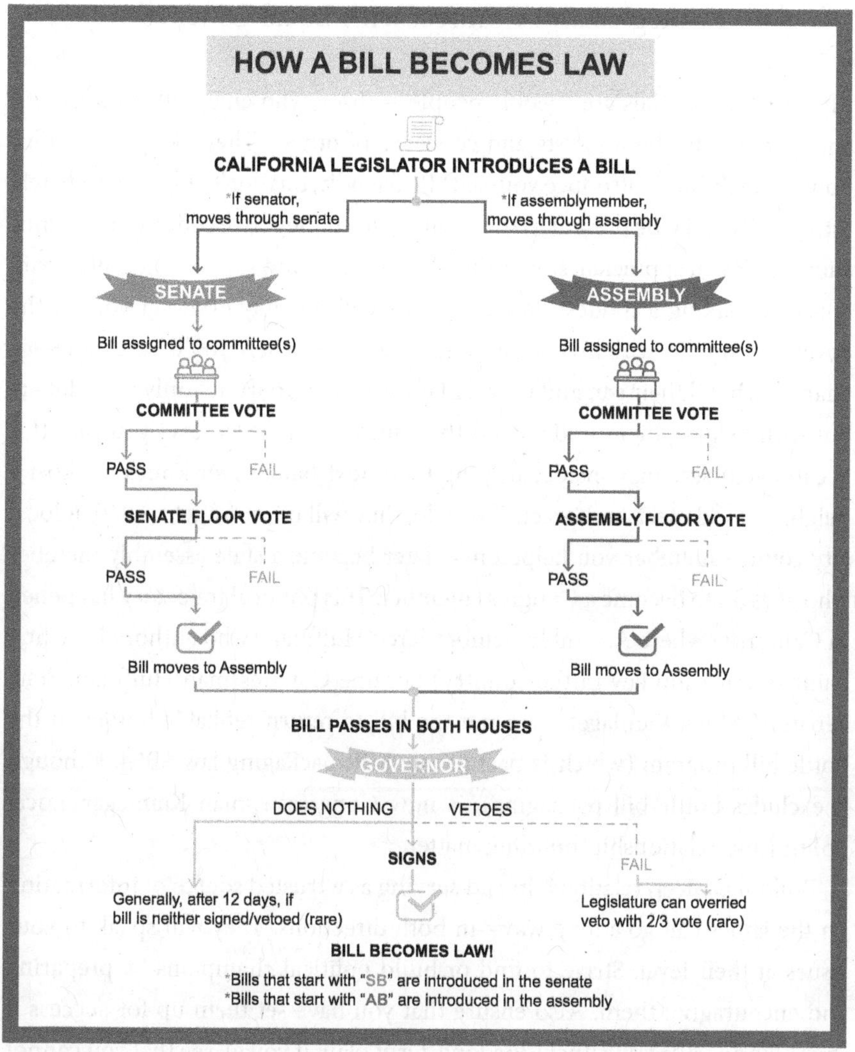

Figure 3.1 How a state bill becomes law.

Source: Courtesy of California State Parks, State Capitol Museum.

recent concerns. Has a consumer product impacted health, safety, or the environment? Have there been wildlife impacts from plastics and other waste? Is illegal dumping or disposal of needles occurring in the area? Perhaps there was a toxic spill recently. It's important to be aware of responses by the community and elected officials.

Strategize and Connect with Local Officials

Most elected officials are "people people"—those who enjoy interacting with and listening to the interests and concerns of others. They likely have office hours. Stop by and introduce yourself! Be friendly, but don't jump immediately into your agenda. Be present in the community. If they have community events, show up. If you appreciate something they've done and support their platform, consider making a modest donation. This will possibly even get you on the coveted donor list with invitations to special events where you'll have more of a chance to bend their ear, and they'll likely listen. Support not only with dollars but with helping them understand the issues, at whatever level you can. If it is campaign season, consider helping them text-bank, door knock, or host a neighborhood meet-and-greet. If you do, they will never forget you! That local city council member you helped may later become a state assembly member who goes on to become a Congress member. This political trajectory happened in California where Assemblymember Jared Huffman (who authored the first paint stewardship law in the country) became Congressman Huffman. State Senator Sydney Kamlager, who ran the bill to return refillable bottles in the bottle bill program (which is part of our EPR packaging law SB54, although it excludes bottle bill packaging), is now Congresswoman Kamlager. Local politics and relationship-building matter.

Building a deep relationship and serving as a trusted source of information on the issues can go a long way—in both directions. They will speak to your issues at their level. Strive to find or build political champions by preparing and encouraging them. Also ensure that you have set them up for success if they take on your issue. Include a long-term plan, if you all see that you cannot pass the bill the first year it is introduced. Just as you don't want to set anyone up for failure, the same holds with elected officials. You want to help your strongest allies avoid failure and vulnerability to their elected position. This benefits them and their stakeholders.

In carefully designing a political strategy, it's helpful to remember that party staff members are critical as well, especially since they're the ones to whom busy elected officials delegate work. It's often a staff member who drafts new legislation, summarizes proposed legislation provided by others, and takes meetings with constituents. It's good to get to know these folks, even outside of the office; they are your allies or should be. Some of them will even care deeply about the issues that you're championing.

If this is an issue that impacts local solid waste operations, as most EPR issues do, obtain statements of support from local officials. If nearby jurisdictions share your problems, bring them into the fold; regional action can be powerful. Remember, too, there are other stakeholders, including unlikely bedfellows, who can take you leaps and bounds beyond expectations. In our dealing with pharmaceutical and needle producers, there were varied players in support of EPR legislation. Included were wastewater agencies, drug abuse prevention advocates, local governments, municipal waste haulers, and HHW program operators whose workers were getting stuck with needles on the materials sorting line. The friends you make in this battle will remember you in the next one.

Designing Effective Strategies

In designing an effective strategy, it helps to take the broader playing field into account, such as affected industries. Look for industry allies. Depending on your approach, your intended breadth of reach may be local, statewide, or national. Determine if the industry associations support circular economy solutions. Are there website claims that they're concerned about the environment? Research can provide some insight into operations and aspirations.

Having a deep understanding of your playing field, and the players on it, is key. In essence, you're deconstructing before you construct. An ability to convey this understanding to the bill author gives them confidence that you are a good bill sponsor. A great author needs a strong bill sponsor.

Present a solution: One of the first lessons many of us learn at work is, "Don't just present problems, tell us how to fix them." If you tell elected officials that a product is a problem at end-of-life and the industry needs to recycle it, they are likely to ask, "Recycle it into what?" If you're talking about carpets or mattresses, don't hold a meeting to discuss producer take-back opportunities until you know the product composition. For those who run landfills, among their biggest headaches are likely carpet and mattresses. Carpet and mattresses are big, bulky, difficult to handle, and take up a lot of room. While carpet may be handy for padding the bottom of a new landfill cell to keep things from piercing the liner, there's far too much carpet to be used that way. Yet truckload after truckload shows up at landfills every day. Carpets as we know are made of fiber (mostly synthetic), glue, and a backing that is made of synthetic jute, fleece, polypropylene, and latex, usually in the form of ground calcium carbonate (PC4) for low oil absorption. Can't something be done with all that? Yes! Most of it anyway. There are manufacturers making products out of old carpet that range from new carpet to floor mats, to auto parts and building materials. California took the lead in passing the first carpet product stewardship bill in the country. To aid the effort, we got creative by bringing manufacturers to meetings, giving factory tours of their facility to officials, and making a short film about their efforts to share more broadly. California is followed by New York in implementing EPR for carpet, requiring producers to fund its collection and recycling.

Negotiate

We don't like to give away things when others don't. A strategic approach involves being prepared to give only that which you can afford to lose. Starting negotiation with your bottom line is not a recommended strategy. Producers and communities benefit from reaching across the table to settle on strong EPR solutions. We have learned enough lessons now to show that there are great risks if we pass bills that do not have strong enforcement tools, oversight agencies, or performance and transparency standards. Those are things that should not be compromised. A "low bar" stewardship bill, like the California carpet bill passed in 2010, was the best we could get at the time with the understanding the legislators had about stewardship programs and the effort the carpet industry would invest in pushing back. We can use this as a lesson: compromise is fine to get a bill that is a great improvement over what you have at the moment. However, if you give producers too much control without good enforcement tools and funded agency oversight, a big problem can arise. It might be best to wait for a better political scenario or "go local" where you can get a better ordinance passed first, to set the tone for the state bill.

These are our three assertions that help us decide whether to undertake the effort to introduce and pass a bill. We need to be relatively certain that

1. It will work.
2. It can pass.
3. It won't get in the way of future progress.

It takes a lot of energy and resources to pass a bill, so we all need to be very thoughtful. But don't let people say you cannot do it and just walk away. If you believe it's important, then do not be afraid to take on hard EPR bills. Just be ready! Then, once you start, do not stop until it passes, no matter how long it takes. Alameda passed the first EPR ordinance for pharmaceuticals in 2012; it was challenged all the way up to the US Supreme Court and we won in 2015. Then SB 212 was passed in 2018 for pharmaceuticals and sharps EPR—a great bill. Now we are implementing it, which means it was twelve years from the time we started meeting to develop the Alameda ordinance in 2010, until it was put into action for Californians. It was hard, but worth doing.

Others are coming much faster now. We passed battery EPR in 2022 that also started in 2012 and SB 54, a packaging EPR that was started only four years earlier but is by far the largest EPR program with thousands of packaging producers and stakeholders. California has 12 percent of the US market share and so companies tend to fight here a lot harder than in smaller states. That means that once it's done in California, chances are that companies want rule harmonization, and it will be easier for the rest of the states, with shorter timeframes required to pass the EPR bills.

Lessons Learned

Know Your Audience and Speak Their Language

If you want to see people's eyes glaze over, fill your sentences with acronyms, technical jargon, and cool-sounding slogans. "By enacting EPR, we can reduce

GHGs, lessen the impact on MRFs, enact IPCC standards, and take a giant step toward carbon neutrality and saving the planet." You may realize that your audience has little idea about what you've said, and less of an inclination to support you.

We are often asked to speak at events, on the radio, and on television. Often, we're paired with other people whom interviewers think will make for a lively conversation. I [Tim] remember a radio interview not too long ago where the other guest was a woman who worked on marine conservation, which is a great crossover topic. I went on and on about all the ways EPR could not only reduce waste and save energy but help clean up the oceans, protect the whales, and so on. When I finally paused for breath, the nice woman asked, "Excuse me, what's EPR?" Professionals, no matter what field, need to take a step back and remember that others don't "live" our topics. They are on a spectrum of differing levels of awareness, knowledge, acceptance, and intention; we need to first gauge where they are and meet them there.

They say that part of Shakespeare's genius was that he could write a play that was as deeply satisfying to the nobility as it was to the penny-paying, standing-room-only audience on the floor.[14] We can't all be Shakespeare, but that's the approach you want to aspire to. Ask yourself, "How can I present this in the brief time I have and in a way that will make sense to the corporate lobbyist, the senator and their legislative director, the environmental advocate, and the bureaucrat who has to make all this work?" Figure that out, and you're halfway to winning the battle.

If you are a policy advocate, you probably know your field better than almost anyone. You've read almost everything, including all those relatively boring technical reports. You may have even written your share of them. The jargon and acronyms fly fast and loose because everyone in *your* room knows them too. Nothing is wrong with that. It's efficient, saves time, and builds cohesion—until it doesn't. It's wonderful to work with a community of like-minded folks, but if you want to get anything done, you're going to have to widen your gaze and talk to people who may have never heard of any of this, or if they have, may not know a lot or may be actively hostile to the idea. It's best

14 "Who Were These People? Audiences in Shakespeare's Day." *Seattle Shakespeare,* https://www.seattleshakespeare.org/who-were-these-people (accessed 2022).

not to assume that the shared values of your small group are shared by anyone else. (You know what they say about people who assume.) Think back to when all of this was new to you. Remember each of the epiphanies along the way? They don't come easily. You may have had to learn a lot of new information and *unlearn* a lot of the ways you were used to looking at the world. It helps in situations like this to adopt what the Buddhists call, "beginner's mind." Share as if it's new to you, keep it simple, connect the dots, and explain everything in a way that will make sense to your audience. Just make sure not to oversimplify or overexplain, so that you avoid inadvertently insulting your audience.

When working on the issue of mercury thermostats, the problem looks different from different vantage points: those of manufacturers, in the construction trades, or as operators of household hazardous waste (HHW) facilities. It's helpful to get them all to the table and start with establishing common goals and vernacular. Your aim is to get them to listen to *each other.* Learning to see things from someone else's perspective can go a long way in helping to solve the problem.

It is optimistic to believe that everyone comes to the table in good faith, but it's crucial to treat them as if they do. We all have our jobs to do, our roles to play; in the end, most of us care about doing the right thing. The high-powered executives and lobbyists from the pharmaceutical companies didn't seem to care too much about what environmental advocates thought of them. They appeared to listen more thoughtfully to the district attorneys, law enforcement officers, pharmacists, and healthcare providers. But it became clear from the variety of discussions that there was a real problem they could help to address. Treating people with respect and respectfully listening can be contagious. Once someone understands that they can work with you despite your differences, you're on your way. You can disagree, but never be disagreeable.

Seek Win/Win/Win Solutions

Abraham Lincoln said, "Do I not destroy my enemies when I make them my friends?"[15] In some ways, the struggle with the pharmaceutical industry was a good example of this in action. We did not demonize the industry, rather

15 Kathleen Foley, "'Lincoln' Seeks to Set the Facts Straight." *Los Angeles Times,* April 3, 1996.

we focused on solving problems. In the end, we made friends, and in other quarters, earned a measure of grudging respect. Heidi was hugged by a lobbyist (for a sharps manufacturer) after the last hearing wherein the pharmaceutical EPR bill was approved to go to the governor's desk. Was it our imagination that he looked around before giving her a hug to make sure that his colleagues wouldn't see this? It pays to avoid offending people; they might just become your allies in passing a bill.

Pharmacists and drug store chains are only too aware of the problems we discussed earlier, and they were happy to seek solutions, especially those within reach and not at their expense. They appreciated the practical approach to writing the regulations and implementing the programs that accommodated shared concerns (e.g., safety, liability, convenience). That approach has won EPR movement allies in some surprising places. We have productive working relationships across many industries, from power providers to carpet and mattress manufacturers, packaging, and many more. It used to be common for chambers of commerce and industry associations to reflexively oppose any regulations. That has changed. We're not best friends, but they know we'll be respectful, we'll listen, and try to find solutions that work for everyone.

Approach discussions with humility, recognizing that in any industry the people know more about their business than you ever will. You might go in with reams of research, great ideas, and the best intentions, but almost always they will point out practical problems that arise because there was something you didn't know. That's okay. There's always more to learn and you don't know what you don't know. Being open to learning and to the ideas of the regulated industry will go a long way to their acceptance of you, as someone they can work with. Remember: A lot of these folks you will see again.

Case in point: In California, environmental advocates, local governments, waste haulers, and many more stakeholders wanted to hold the plastics and packaging industry accountable for what they were putting in the marketplace. They attempted to pass companion bills in the state's assembly and senate starting in 2018, but both failed in the two-year session.[16] The environmental

16 Clare Goldsberry, "California Legislature Fails to Vote on Plastic Packaging Bills as Session Ends." *Plastics Today*, September 1, 2020.

advocates were fed up so they pooled their resources and qualified a ballot initiative that would put a tax on plastic producers to fund environmental improvements, require plastic source reduction, and ban polystyrene. The problem was that the industry despised the ballot measure, and representatives were gearing up a nine-digit opposition spending campaign that kicked-off with a full-page ad in the *Los Angeles Times*. Neither the industry nor the environmental community wanted the uncertainty of a ballot measure, so they chose to work together; the bill was a win-win-win solution for both supporters and prior opponents, but also the public, who did not have to wade through ads all fall to figure out how to vote on the ballot measure on such a complicated topic of packaging waste reduction.

Our team spent ten months on calls, reviewing bill drafts, and in online and in-person meetings negotiating the details of the deal, which resulted in the passage of SB 54 (Allen) on June 30, 2022, the country's most comprehensive EPR law to date.[17]

SB 54 (Senator Ben Allen): The EPR Lessons Learned, All in One Bill, Plus More!

Heidi calls SB 54 "the big kahuna" because packaging EPR is the most complicated EPR bill that can be done, for many reasons: the number of producers and stakeholders, materials, and various infrastructure across a state to collect and process it all.

So how did we meet the challenge and pass such a bill? It took over a decade of hard work educating local governments, state officials, haulers, recyclers, and municipalities about EPR and bringing them on board. We passed EPR bills with those stakeholders, sometimes also working together on "clean up" bills to ensure the lessons learned are in the legislation.

Another reason, we believe, is because of Heidi's foresight in taking legislators on annual study trips to Canada with the California Foundation

[17] Lindsey Parkinson, "California Passes Strongest EPR and Recycling Legislation in the US." *Food Packaging Forum*, July 5, 2022.

Table 3.1 Legislative EPR in California

Year	EPR Bills and CE Bill by Material	Bill Number
2024	Textiles	SB 707
2024	PaintCare Expansion Act to include: All Nonindustrial Paint and Paint Products	SB 1143
2023	Aerosol Coatings	AB 1526
2022	Loose Batteries	AB 2440
2022	Single-use Packaging and Single-use Plastic Food Service Ware	SB 54–Circular Economy Bill
2018	Medications and Sharps	SB 212
2013	Mattresses	SB 254
2010	Carpet	AB 2398
2010	Architectural Paint	AB 1343
2008	Ag Pesticide Containers	SB 1723
2008	Recalled Products	AB 1860
2008	Mercury Thermostats	AB 2347

Source: National Stewardship Action Council, 2024.

for the Environment and the Economy (CFEE), to see EPR programs up close beginning in 2018. Those trips developed legislators' understanding about what was possible, while also allowing travelers to develop relationships and trust. That mattered, especially when it came time to pass a tough bill like SB 54.

So, what is the total list of stewardship or EPR bills that have passed in California that provided valuable lessons to develop SB 54? See Table 3.1.

There is a lot that is very good about SB 54 (Allen). A forty-one-page bill, SB 54 filled the gaps that we were aware of in other EPR bills that had passed. The most exciting aspect to us is that it is the world's first true circular economy bill. By that we mean that it covers three main aspects of a circular economy. It aims to (1) reduce waste and pollution, (2) keep materials in motion, and (3) regenerate natural systems.

Key bill provisions that address these three pillars are:

1. Reduces Waste: Plastic packaging must be source reduced by 25 percent by 2032 by weight and by unit. Of that 25 percent, 2 percent must be through moving to refillable packaging by 2027 and 4 percent by 2030. Ten percent of plastic packaging must be source reduced through shifting to refillable-reusable packaging or elimination. The remaining shall be

source reduced through concentration, right-sizing, light-weighting, shifting to bulk or large-format packaging, or shifting to nonplastic material.

2. Keeps Materials in Motion: All packaging must be recyclable or compostable by 2032. Recycling targets are 30 percent by 2028, 40 percent by 2030, and 65 percent by 2032. Expanded Polystyrene (EPS) must hit 25 percent recycling rate by 2025 or it cannot be sold in California.
3. Regenerates Natural Systems: Creates a "Pollution Mitigation Fund" that packaging producers must fund $500 million a year for ten years totaling $5 billion used to clean up California with 60 percent going to environmental justice and public health efforts.[18]

Another reason to love SB54 is that the process was right. The drafting of the bill was very well thought out and developed over ten months, with almost every stakeholder represented and buy-in and support from all the key stakeholder groups—a very important aspect of the bill for the program to work.

Finally, there is substantial language in the bill to protect under-resourced and rural communities, the same communities in which many waste management facilities are sited. It is impressive, and unusual, to see a bill to be so equitable and protective of communities that have traditionally borne a disproportionate burden of pollution. In addition to providing protection for these communities, SB54 protects these communities and helps to repair pollution damage through the Mitigation Fund (described earlier in #3 of Key bill provisions).

(To view the NSAC and Recycling Partnership webinar about EPR and the SB54 bill, presented by people who helped draft and pass the bill, please visit https://www.youtube.com/watch?v=09oDM6dS-Zo.)

Remember That "The Devil Is in the Details"

Comparing some of our first EPR bills to more recent ones, it's tempting to laugh or cringe. There was so much we didn't know, we made mistakes, and

[18] Plastic Pollution Prevention and Packaging Producer Responsibility Act (2022) Section 42064 PRC.

we couldn't convince legislators to pass bolder bills. When the details are not right, they become problematic. Bills now are much longer and more detailed to incorporate all that we learned in prior EPR bills. Holes have been plugged. Best of all, there are processes in place for problems we still likely don't anticipate, because they will always happen to varying degrees.

Industries engaged in EPR set up a producer responsibility organization (PRO) to assume the responsibility for essentially running the EPR program wherein producers are held accountable for their products at end-of-life. It makes sense that PROs are 501(c)(3) nonprofit organizations, although this designation is still in debate in some circles. After all, they are not there to make money but to provide a "public benefit" which is a requirement of a charitable 501(c)(3). But did you know there are twenty-one different kinds of nonprofits in the US tax code? We didn't. That is, until we ran into some issues with PROs that weren't behaving as we had intended in the law or ordinance. They did exactly what the law allowed but we didn't foresee the problem. If you work for an environmental group, your organization is probably a 501(c)(3) or 501(c)(4). Such groups are set up for the benefit of the public. Chambers of commerce, industry associations, and similar groups are usually organized as a 501(c)(6). The crucial difference is that these groups exist to promote business, for the benefit of *their members*. They aren't set up to generate a profit and their earnings don't benefit shareholders, although they can participate in revenue-generating events and spend money on lobbying. They apparently can even use it to file a suit to cancel your program, which is what happened to us in California. So now, one of the requirements that CalRecycle puts into every EPR law is to require that the PRO is a 501(c)(3) nonprofit that must operate for "public benefit," not "association benefit" or "mutual benefit." We understand now that you can prevent a PRO from being able to use the consumer fees or program funds to sue the oversight agency. Being on the cutting edge can come with a price, but you also get to make mistakes from which others can learn. When doing your homework and learning all that you can, include seeking help from those who have been down the EPR road before. It pays to be higher on the learning curve.

Defining Success

Every EPR campaign so far has resulted in additional lessons learned. One is to start with asking yourself questions: How do you define success? What is a sufficient degree of effort? How convenient must the program be for the public? When there's sufficient effort, the program is convenient, and an EPR bill is passed, are you done? Was it successful?

In the long, drawn-out negotiations over California's medication and needle EPR bill, it was finally agreed that the industry must establish at least five collection locations in each county. The pharmaceutical industry was happy to leave it there, but we battled to add a provision that allows an eligible site that wants a bin to get one within ninety days of asking. This means that they are obligated to give a safe medication collection receptacle to every eligible pharmacy, hospital, and law enforcement location that asks for one, even if they have a collection site next door. Adequate infrastructure is an essential component of success.

Someone in government will be overseeing your new EPR program; they will need staffing and will have other expenses. These resource costs are usually built into strong EPR legislation. Often a new program is set up with a significant lag time to allow for planning and preparation. But if the revenue doesn't begin to flow until the program is fully active, regulatory agencies struggle to cover these costs. We have learned to require revenue to begin flowing as soon as possible, to enable regulators to do their job and avoid delay by regulated industries.

Remember that saying about the fox guarding the hen house? If the industry being regulated is the one (potentially begrudgingly) operating the program, there can be, and have been, some issues. As a result, there must either be an advisory committee put in place with the power to stop a program or ensure that the right stakeholders are on the PRO board so that adequate decisions are being made and for the right reasons. The PRO must listen to the advisory board and if they do not, they should be held accountable to provide sufficient explanations as to why they are not.

How Will You Know If Your EPR Program Is a Success?

We have found it effective to specify measurable goals in the legislation, whether in tonnages, percentages, or otherwise, along with providing target dates ("rates and dates"). These basic stipulations are a requirement for any bill that we support, or at least ensure the oversight agency is solid and can set those goals in regulations. Plans and promises do not suffice. We learned this the hard way when a carpet industry bill in 2010 had "continuous and meaningful improvement" as a measurement of the program's success. If there is room for argument about whether a program is meeting its goals, there will almost certainly be one. Adopt SMART criteria (Specific, Measurable, Achievable, Relevant, Time-bound) and you are likely to avoid this problem. It seems to work best to have multiple annual benchmarks, allowing for corrections to be made along the way as needed.

For EPR to be successful, there must be accountability and significant penalties for noncompliance or failure to meet goals. If you're trying to get a multi-billion-dollar industry to change course, expect it to be tough. If they can avoid changing by paying a nuisance fee every year, they likely will do that. In the California Beverage Container Recycling Bill ("bottle bill") example mentioned earlier, where consumers pay a deposit for the beverage container that can be taken to a redemption location to receive their deposit (e.g., 10 cents), could work more effectively. Grocery store owners, for instance, can pay an insignificant penalty to be exempted from serving as a beverage container collection point, and locations where you can redeem the container are few and far between. This essentially makes that extra deposit fee a "tax" because it's likely that most containers will not be returned and thus, not redeemed for cash. The reason for this? For grocery store owners to provide container collection (such as a designated area with reverse vending machines), the fines would need to be significantly larger. A more proactive and positive approach is an incentive, such as a program where beverage producers pay the retailers for the space and labor to take back the beverage containers. This is done in bottle bill programs run by producers in British Columbia but not yet in California's government-run program.

And there must be a process for pulling the plug. Sometimes, despite the best intentions and great effort, some programs just don't work. Be sure you have a process for making this determination. For example, the mercury thermostat EPR program in California, created by a bill passed in 2007, was fraught with noncompliance and little effective oversight. In 2021, NSAC negotiated with the mercury thermostat industry on AB 707 (Quirk), a bill that was signed into law by Governor Gavin Newsom that year and which made significant improvements to the program. Improvements included removing the existing PRO as program operator in favor of a third-party operator and establishing the financial incentive for product take-back at $30 per mercury thermostat returned.

The PRO has control. After passing two bills in 2017 and 2019 to improve the existing carpet EPR program in California, we learned that CalRecycle did not have enough authority to "take the keys away" from the PRO. This was especially difficult because the "hammer" or penalty in the carpet stewardship program example was the banning of carpet sales in California. This was unrealistic and the carpet industry knew it.

To compound matters, money from consumer fee collection was held in a bank in Georgia, six states and two time zones away. CalRecycle (CIWMB at the time) wanted internalized costs where producers pay the PRO directly and must answer to their shareholders what those product end-of-life costs are for so they will learn it's cheaper to "design for the environment." This was preferred over consumer stewardship fees. While the political environment would not allow us to eliminate the consumer fee, we realized then that the money needed to be kept somewhere in California where it could be protected from individual PRO control and ensure it was used for future implementation of the law. Lessons like this make every new bill stronger.

In many cases, passing a bill or ordinance is not the end of the process. Recently, it has become common for bills and ordinances to contain a program outline and a directive to regulators to craft a program that fits. Our work is not done at this point; it's critical to stay engaged. The most well-intended program can go off the rails by the time industry representatives are done adding exceptions, exemptions, delays, or reduced penalties. Unless we are there to keep things on track and push back as needed, this is very likely. Even after, when the regulations are complete and the program is launched, you'll want to read those annual progress reports, check the data, weigh in and encourage enforcement by the appropriate agency, and more. This is an iterative process. Getting involved means having to stay involved; the battle is not won in one day or a year.

Celebrate Your Wins, Learn from Your Mistakes, Share Lessons Learned

The National Stewardship Action Council (NSAC) holds a legislative leadership awards ceremony every year, and smaller events throughout the year. We give plaque awards to our heroes, elected officials, regulators, courageous industry innovators, and others. We acknowledge how tough it is to get these things done and how much blood, sweat, and tears go into the process. It is all well-deserved and heartfelt, provides public recognition, and confirms commitment to the issues. We sound the trumpet far and wide for these accomplishments including in the press, on our website, at conferences, and everywhere we can. You want your allies to be appreciated and publicly applauded for their commitments and efforts, especially for taking on hard issues. Additional benefits come from this such as cementing alliances, fostering positive feelings about our organization, and smoothing the path somewhat for the next challenge.

While every battle, of course, doesn't end in victory, NSAC has had a good run so far. We've lost some at the last minute when it seemed that victory was within our grasp. Sometimes, the result is an unnecessary surprise, such as when an elected official becomes perturbed over a perceived slight, decides to skip a crucial vote, and your bill vote goes down by one. Or it just got lost in the shuffle because of too many bills and a staffer forgot to put it on the calendar. Such is life. Take a minute to detail with dismay, comfort each other, and then dust ourselves off. But don't quit.

Some EPR bills seem to just fall into place in many California communities, be it due to the amazing people, culture, timing, precedence, and aligned stars.

For rechargeable battery EPR, an industry-funded group offered to set it up and soon there were battery drop-boxes in many hardware stores, libraries, and other community locations. It was easy, simple, and relatively cost-free.

Sometimes though, you just need to start over. A packer truck was destroyed by a fire started by a lithium-ion battery. The material recovery facility (MRF) went up in flames next, and the dominoes continued to fall. Insurance companies, shipping companies, and others reacted quickly. Insurance companies increased the number of policies and their costs for coverage, with the threat of cancellation should the fires persist. The PRO that ran the program announced new guidelines out of desperation. These included requiring anyone participating in the program to complete online safety training; that all batteries be hand-sorted before shipping and bagged individually, or that their terminals be carefully taped; and that rechargeable batteries be separated from other types of batteries. The almost universal response from busy librarians and retail workers was, "No thanks," and so the program withered. Only now, years later, is California tackling statewide battery EPR, having learned important lessons. Rather than ask already busy employees to become battery sorting and recycling experts, we are requiring the industry to set up the system and compensate everyone for their time.

There is a lot of work involved in product stewardship and EPR, but you don't have to learn from scratch the way we did. Take advantage of the lessons learned—especially those learned the hard way. Remember that every place and everyone is different. Do your homework, find your allies, and carefully develop a thoughtful strategy based on insights. Ask others for help if you need it. EPR and the circular economy are no longer new ideas. There are more and more successes every day that you can point to and say, "See? It works!"

Summary

In this chapter, we've shared many of the things we have learned through our decades of working in this field and on EPR. From our experiences, real examples were provided to, perhaps, bring readers higher on the learning curve. There are strategies and tools to consider and possibly employ, recognizing that what worked for some, may not work for others; adapt to suit your own needs. We have hopefully prompted thoughts and questions for readers to consider and answer that help to draft, introduce, pass, and implement a best-in-class EPR bill. From gathering data to making a case as to why there should be a law; developing the resources, relationships, partnerships; getting the timing right; and working to ensure that the EPR bill has the details that matter most. We hope that by sharing information and insights, we have provided some assurances and uncovered unforeseen matters to help build confidence in taking steps toward EPR legislation. This drives us all closer to a more equitable, circular economy. We need to increase the momentum for change: we have already done so much damage to our beautiful planet and our shared resources, that there is no time to waste. Our individual and collective actions matter now, and even more so for the generations that follow. Good luck on your EPR journey.

4

Mattress Management

Part One: Mattress Management through Industry Engagement and Stewardship: How It Works

Mattress Recycling Council

Mattresses are large, bulky, somewhat lightweight, durable consumer products upon which we all depend. These traits also make them problematic to dispose of at the end of their useful life. Our options are to properly dispose of our unwanted mattress by bringing it to a landfill, transfer station, or mattress collection site or event. While mattresses are typically less than 1 percent of the municipal solid waste stream, their bulk, construction, and resiliency mean they occupy a disproportionate amount of space in solid waste transport vehicles and landfills. They can also be problematic: mattresses with springs can become tangled in landfill compactor equipment and incinerator feed grinders often pulling solid waste disposal equipment offline for a period of time. The good news is that depending on material quality and market conditions, many mattress components (e.g., steel wire, foam, cotton) can be recycled.

To better understand the fate of unwanted mattresses, the International Sleep Products Association (ISPA)—the trade organization for the mattress industry—assembled the Disposal Task Force and surveyed mattress retailers operating across the United States in 1991 to determine what was happening to bedding at the end of its useful life. Retailers said they sent used bed sets to landfills, gave all or some of the used bedding to charities, or sent old bed sets to renovators.

In 1992, ISPA sponsored an experimental program in Baltimore, Maryland, to turn old mattresses and foundations into energy. Through the heavily promoted test, which had the financial support of ten brand-name manufacturers as well as other groups, they hoped to convert 10,000 mattresses into tens of thousands of volts of energy from the city's waste-to-energy facility. Only 800 mattresses were collected for this effort; the industry viewed the project as a stepping stone, not a stopping point.

In 1994, a new project was launched when the US Department of Justice's Federal Prison Industries agreed to work with the industry on a joint mattress recycling project. Prisoners at a federal correctional institution in Fairton, New Jersey, dismantled mattresses collected by area retailers. However, there was a lack of markets for the used mattress components. This scuttled the prison project and stymied industry recycling efforts for the decade to come.

While some companies pursued their own initiatives, several factors made the economics of recycling more attractive. For one, limited landfill capacity and increased demand drove up tipping fees that are paid to dispose of waste at a landfill. In response, some governments began banning entire classes of products from landfills, including mattresses. In addition, mattress recyclers began to spring up across the United States and Canada. Some of the earliest were created by charities that sought jobs for populations that had trouble finding and retaining employment such as those with some types of disability, a criminal record and prior incarceration, or experiencing homelessness. Throughout the 2000s, new nonprofit and for-profit recyclers opened or expanded their operations. By 2015, there were more than fifty recyclers processing mattresses in the United States.

In the early 2010s, ISPA advocated, unsuccessfully, for federal mattress recycling legislation. Based on that experience, ISPA learned that Congress viewed recycling as largely a state or local issue and shifted its focus to crafting statewide solutions.

In 2013, Connecticut became the first state to establish comprehensive mattress recycling stewardship legislation (Public Act 13–42). Part two of this chapter, authored by Tom Metzner of CT DEEP, further highlights the Connecticut mattress recycling experience. Along with Connecticut, Rhode Island (General Law 23–90) and California (Public Resources Code sections 42985–42994) enacted EPR laws for mattresses, also in 2013. In 2022, Oregon adopted EPR for mattress recycling (Senate Bill 1576).

To implement these laws, ISPA, representing mattress manufacturers, created the Mattress Recycling Council Inc. (MRC), a 501(c)(3) nonprofit organization. Each state's law requires that retailers (including those online) and other sellers collect a recycling fee on every mattress and foundation they sell to consumers and other mattress purchasers in that state. The law also requires that they itemize the fee on the invoice or sales receipt provided to the purchaser, and remit the fees collected to MRC, which in turn uses that money to fund the recycling efforts. The laws additionally require MRC to educate consumers and the mattress industry about the recycling program requirements.

The laws in each state are not identical, but they are broadly consistent with each other. Each state law requires that MRC develop and submit a comprehensive program plan to state regulators for their approval. These program plans address criteria set in each state's law and provide a detailed description of how MRC will administer the state's mattress recycling program.

During the plan development process, implementation, and afterwards, MRC closely collaborates with state and local governments, solid waste service providers, mattress recyclers, transporters, mattress retailers, and environmental organizations. In addition to engaging with these parties, each law requires MRC to submit to state regulators a public annual report describing its accomplishments, and for the state agency to exercise regulatory oversight regarding MRC's activities and performance.

The extent and type of regulatory oversight varies from state to state. For example, a state may collect mattress-discard data from parties other than MRC and recyclers to measure a program's overall effectiveness and identify areas for improvement, set quantitative performance goals (such as unit collection and recycling targets) for its program, or use its enforcement authority to compel compliance with the law.

How It Works

Collection Network

The average age of mattresses and foundations received by MRC's programs is 13.9 years.[1] In each state where mattress recycling laws have been enacted and MRC serves as the mattress stewardship organization, MRC works within the existing solid waste collection infrastructure to separate and recycle mattresses collected through those systems. This avoids costly collection redundancies and interference with existing hauling contracts.

MRC's collection network aggregates discarded mattresses from a variety of collection channels including existing bulky item curbside collection programs, its no-cost drop-off network, municipal collection events, and commercial volume sources like retailers and hotels.

Curbside Collection Channels

Waste haulers and municipalities often provide curbside bulky item collection of mattresses to residential customers who schedule a pickup. If collected units are in reasonably good condition, the mattresses are loaded into MRC-provided storage containers positioned at their facility. Solid waste facilities serving as permanent public collection drop-off sites may also place curbside units collected by third-party haulers in MRC-provided storage containers at their facility.

No-Cost Drop-off Network

Most of the permanent networks developed by MRC include existing solid waste facilities (landfills, transfer stations, public works yards) where their mattresses will be consolidated and transported to nearby recyclers. Consumers may also drop off mattresses directly at a recycling facility.

1 Mattress and Foundation Age Analysis. July 2013. Mattress Recycling Council. https://mattressrecyclingcouncil.org/wp-content/uploads/2023/07/2023-MRC-Age-Study-Final-7-24-23-web-posting-2.pdf.

Municipal Collection Events

To supplement the no-cost permanent collection network, MRC hosts mattress collection events where residents can drop off mattresses (also at no cost). MRC partners with municipalities, waste services providers, retailers, and other interested parties to hold collection events where MRC provides labor, transportation, recycling, and publicity. MRC may collaborate with other organizations offering similar collection events to increase community access and participation. While events represent a very small percentage of total units collected, they are often needed to increase convenience for residents. The events also target underserved areas and help to mitigate illegal dumping activity in communities.

Retail and Other Commercial Volume Opportunities

MRC's Commercial Volume Program (CVP) serves mattress retailers, lodging establishments, educational facilities, military bases, and healthcare facilities. Eligible entities with a minimum number of mattress units (set by each state program) can request no-cost transportation to a participating recycling facility. These entities can also choose to drop off units at a designated collection site or a recycler.

Direct pick-up from commercial entities is highly efficient and has environmental and financial benefits to the program because these discarded units tend to be clean, dry and are typically collected in full 53-foot trailer quantities.

To determine where to focus resources and expansion efforts, MRC regularly analyzes residents' access to this collection network.

Transportation

To provide mattress transportation throughout each state, MRC contracts with a mix of third-party logistics companies, transportation companies, mattress recyclers, and solid waste facilities to transport mattresses from the collection network. MRC coordinates the transport of over 15,000 loads of mattresses annually.

Because mattresses are bulky and light cargo, creating efficient transportation networks is important to control program costs. MRC pays for storage containers that are appropriate for the size and location of each facility, which may include 53-foot trailers, sea containers, and roll-off bins for sites with limited space or other logistical constraints. These 53-foot trailers are preferred because they hold the most mattress units, resulting in a lower per-unit transport cost. Smaller roll-off containers are the least efficient alternative but may be the only option for collection sites with small or restricted loading areas. All containers provided by MRC are covered to reduce contamination of mattresses and keep them dry.

Mattress Recycling

MRC contracts with third-party recycling facilities to dismantle old mattresses and box springs from commercial and residential generators. MRC recyclers maximize the recovery of mattress components, striving to recycle at least 75 percent by weight of each mattress processed through diversion of resources like steel, foam, and wood from landfills. MRC educates staff at collection locations on suitability requirements. When mattresses arrive at a recycling facility, the load is evaluated for bedbugs and for heavily soiled units. To minimize facility contamination and worker exposure, mattress units that can't be safely recycled are sent to the landfill, although it's somewhat rare that a unit is too contaminated to be recycled.

Dismantling methods vary from one recycler to the next, but most use a combination of manual and mechanical processes. Each recyclable material type is separated and is often compressed into bales for shipment to buyers. High-quality used mattresses or their component materials may also be sold to renovators, who use them to rebuild mattresses for sale in secondary markets. For materials that can't be sold in scrap markets, MRC encourages its contracted recyclers to explore alternative uses to minimize the amount of material landfilled or incinerated. Figure 4.1 details the most common mattress components.[2]

[2] 2022 California Annual Report, Mattress Recycling Council of California, LLC, June 30, 2023, pp. 33–5. From chrome-extension://efaidnbmnnnibpcajpcglclefindmkaj/https://mattressrecycling

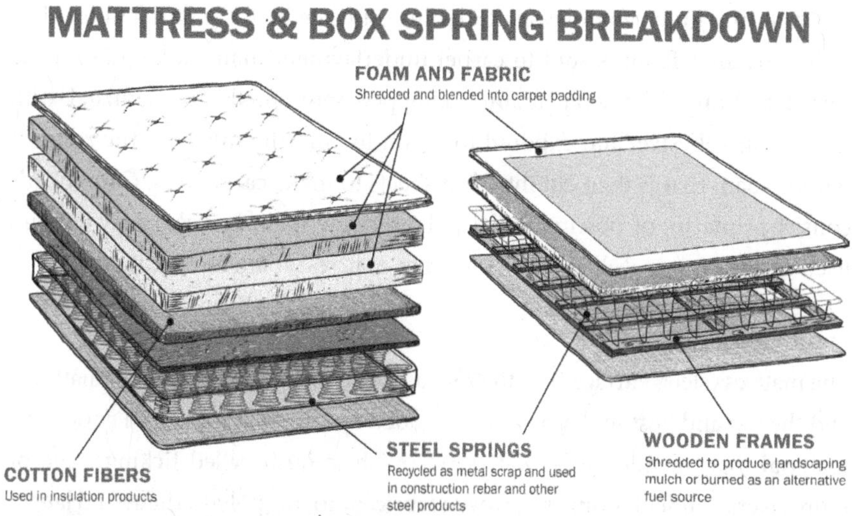

Figure 4.1 Composition of a mattress.

Source: Courtesy of the Mattress Recycling Council.

Steel Wire

Most used mattresses contain steel springs, either in the form of innersprings or pocketed (fabric-encased) coils. Likewise, most used foundations contain a steel wire interior structure. The steel from both mattresses and foundations can be shredded or baled for sale in steel scrap markets. Steel mills then purchase the scrap steel and re-melt it for use in manufacturing new steel products, such as appliances, automobiles, and building products.

Most scrap metal buyers reject pocketed coils for recycling because they consider the polypropylene fabric encasing the coils, which amounts to about 10 percent of the pocketed coil's weight, to be a contaminant. Pocketed coils are best managed by shredding and magnetically separating the metal from the polypropylene sleeve.

Foam

The interiors of most mattresses contain one or more layers of polyurethane and/or latex foam. After the outer quilt panel is removed, the recycler

council.org/wp-content/uploads/2023/06/MRC_annual_report_2022_FINALweb.pdf (accessed June 30, 2023).

separates the inner foam layers for baling and sale to scrap purchasers. Most post-consumer foam is sold to carpet underlayment manufacturers to make carpet padding. The scrap foam is chopped into small pieces, mixed with rebonding adhesive, and exposed to steam to cure the mixture. The result is rebond foam that is then cut into thin sheets to make carpet pad. Carpet pads consist primarily of postindustrial polyurethane foam but also may contain used carpet pad, as well as post-consumer mattress foam and quilt panels.

Quilt Panel

The mattress sleep surface (i.e., the top layer of a one-sided ["no-flip"] mattress, and the top and bottom layers of a two-sided mattress) is typically a composite material that includes a decorative outer fabric (also called ticking), one or more layers of fiber batting, nonwoven fabric, or foam (polyurethane and latex), and a thin backing fabric (usually a nonwoven fabric). These layers usually are quilted together. Once the quilt panel has been removed from the mattress, recyclers typically bale it for sale separately from other reclaimed commodities. Demand for recycled quilt panels is limited for several reasons. First, quilt panels lack consistent construction and input materials. Second, quilt panels are more likely to become contaminated with dirt, water, bodily fluids, or other environmental impurities because they are the outermost layer of a mattress. Quilt panels are most commonly sold to carpet underlayment manufacturers, which chop the quilt into small pieces to make carpet pads. Post-consumer quilt panel represents a small percentage of the finished carpet pad.

Foundation Wood

Most consumers use a foundation or box spring underneath their mattress. The foundation raises the sleep surface of the mattress to a comfortable height and helps prolong the life of the mattress by acting as a shock absorber. A typical foundation is composed of a wood frame assembled using nails or staples. A steel wire structure attached to the wood frame by staples or nails provides the spring support for the foundation. Some foundations contain a wood interior structure instead of a steel wire structure. The sides of the foundation typically are covered by a decorative ticking while the top and bottom typically are covered by a nonwoven fabric.

To dismantle a foundation, recyclers remove the ticking and nonwoven fabric, then shear or shred the unit to separate the metal from the wood. If the wood is free of significant metal or fabric contamination, it can be finely shredded for use as mulch or burned for energy value in a biomass or waste-to-energy facility.

Cotton

Depending on product construction, a mattress may contain other nonwoven fiber layers such as mote cotton which is a by-product of the cotton industry. Fiber markets for used mote cotton are limited and depend on whether the reclaimed material is commingled with other fibers and prevailing market conditions. End uses for this material include insulation and acoustic panel products for the construction and building industry. Cotton may also be managed at municipal composting operations.

Shoddy Felt

Another common layer in mattresses is shoddy felt (also called shoddy pad or just shoddy). This material usually is placed between the metal springs and foam layers to insulate the foam from sharp points on the springs and to keep the foam from working its way into the springs during the life of the mattress. Shoddy is a nonwoven material composed of mixed shredded recycled postindustrial fabric and apparel. Given the mixed nature of the fiber types, this material is usually landfilled.

Other Fiber

Many mattresses also contain layers made from one or more other types of fiber. Depending on the construction of the mattress, these fibers may include polyester, rayon and other synthetic fibers, wool, horse and other animal hair, and coconut fiber. While coconut fiber can be composted if shredded before composting at large compost operations, markets for other fibers are limited. As a result, most are typically landfilled.

Plastics

Retailers often place discarded mattresses they collect from consumers in plastic bags (usually polyethylene) so that dirt, bedbugs, or other biological materials on the discarded units do not contaminate new units being delivered on the same truck. These bags are frequently on the discarded mattress when it arrives at the recycler. Most recyclers bale and sell this plastic to recycling companies who convert the plastic wrap into pellets. Manufacturers of new plastic products use this recycled plastic as a feedstock in their manufacturing processes. Potential end-uses for these pellets include composite plastic lumber and plastic film products.

Cardboard

Mattress foundations may contain a layer of cardboard that separates the wood base structure from the bottom nonwoven fabric. Recyclers bale the cardboard and sell the bales to companies that make chipboard for cereal boxes, corrugated cardboard, and other packaging materials.

Each year, MRC recycling programs on average[3]:

- process nearly 2 million mattresses,
- divert over 65 million pounds of mattress materials from landfills, and
- recycle approximately 75 percent (by weight) of each mattress processed.

Research

A main goal of the MRC is to improve the sustainability of mattress recycling through innovation, collaboration, and communication between suppliers, manufacturers, collectors, dismantlers, and end-market commodity buyers to increase mattress recycling rates. To assist in developing and managing these research efforts, MRC staff works with a team of research and innovation subcontractors to identify projects that have the potential for short-term, high-impact results, as well as longer-term projects with multi-year horizons. From a portfolio management perspective, MRC works to establish and manage diverse projects that explore incremental and disruptive technology development such as mechanical recycling technologies used in adjacent

[3] 2022 California Annual Report, Mattress Recycling Council of California, LLC, June 30, 2023. From chrome-extension://efaidnbmnnnibpcajpcglclefindmkaj/https://mattressrecyclingcouncil.org/wp-content/uploads/2023/06/MRC_annual_report_2022_FINALweb.pdf (accessed June 30, 2023).

industries, process automation innovations, and newly emerging advanced recycling process development.

Many factors influence market acceptance of new technologies. Thus, research is necessary to improve long-term industry performance and shouldn't be focused exclusively on short-term results. To this end, MRC evaluates research proposals, considers proposer qualifications to undertake the project, and assesses the value proposition.

The focus of MRC research projects include:

- Collection: Funding improvements to mattress collection facilities and other operational enhancements can increase landfill diversion and the recyclability of discarded units. These improvements may include paving areas to improve unit quality (by protecting them from dirt and moisture) and decrease mattress loading times, as well as funding collection site equipment that will increase efficiency or improve worker safety when loading discarded units. MRC also collects data to determine which locations are the most and least efficient at loading trailers.
- Transportation: MRC hauls over 15,000 loads of units annually and closely monitors the costs and performance of third-party transporters. By using data collection and analytics, MRC has increased the number of units loaded per container, improved hauling efficiencies, and reduced both transportation costs and the environmental impact of the transport function by reducing the amount of fuel consumed (and associated greenhouse gasses and air pollution generated).
- Recycling: MRC needs efficient recyclers and offers Lean Manufacturing training to help California recyclers to eliminate operational inefficiencies and increase productivity and profitability. Participating facilities receive a confidential report detailing the results of the analysis and recommendations for improvements. These assessments have shown that most recyclers have the potential to increase their productivity by an average of 30 percent by reducing non-value-added activities like excessive forklift movements and double handling of materials.
- MRC also provides fire prevention assessments to recyclers to help them minimize fire risks by identifying and preventing high-risk activities and conditions.

Figure 4.2 Environmental benefits of MRC's California program.

Source: Courtesy of the Mattress Recycling Council.

- End Markets: MRC recognizes that as its activities expand, the post-consumer materials that its recyclers generate may overwhelm existing markets for those commodities. Demand for some materials (such as post-consumer foam used to make carpet pads) may decline as carpet share of the floor-covering market diminishes. For these reasons, MRC has sponsored a variety of research projects to develop new and better end markets for mattress commodities. Each of these projects is posted on the MRC website (www.mattressrecyclingcouncil.org).
- Life Cycle Analysis: In 2021, MRC conducted a life cycle analysis (LCA) of its California operations to establish baseline environmental performance parameters for the mattress recycling industry. The LCA found that the MRC mattress recycling program offers significant environmental benefits (see Figure 4.2). MRC expects to use the results of this study as a benchmark for evaluating future recycling technologies.

Education and Outreach

MRC's education and outreach efforts target a broad range of audiences, including the mattress industry, state and local governments, the solid waste community, and consumers. To delineate and reach these various audiences, MRC developed the Mattress Recycling Council and Bye Bye Mattress brands (see Figures 4.3 and 4.4).

Figure 4.3 Logo of the Mattress Recycling Council.

Source: Courtesy of the Mattress Recycling Council.

Figure 4.4 Bye Bye Mattress logo.

Source: Courtesy of the Mattress Recycling Council.

For the mattress industry, the MRC brand builds awareness of mattress recycling laws and its reputation as a recycling leader and expert. Today, the brand is widely known in mattress industry publications and is a credible education source for mattress manufacturers, distributors, and retailers about their obligations under each state's mattress recycling law. It communicates with potential and registered participants and develops resources for retailers to communicate the recycling fee and the importance of recycling correctly and effectively. The brand is also responsible for advancing and communicating the importance of sustainability across the mattress supply chain, which is critical to MRC's long-term success. MRC also educates the public sector about efforts to prevent and mitigate illegal mattress dumping.

On the consumer side, MRC developed the Bye Bye Mattress brand, logo, and website (ByeByeMattress.com) to inform residents and businesses disposing of mattresses about recycling and ways to access the program (see Figure 4.4). An online locator allows consumers to search and find nearby mattress recycling locations or upcoming events. The site also features a directory of municipal bulky item programs that offer at least one free pick up per year to residents. Sections of the site are also dedicated to explaining how businesses can access recycling, consumer tips to avoid or prevent illegal dumping and information about the retailer's obligation to offer to take back an old mattress when delivering a new one in California.

To build and support both brands, MRC uses a robust mix of point-of-sale materials, public service announcements, paid advertising campaigns, social media, digital campaigns, direct mail, media outreach, and events. MRC marketing materials and resources are multilingual, and the communication strategy is designed to reach targeted diverse audiences.

MRC annually tracks consumer awareness and conducts periodic research to understand and improve consumer outreach and education. We also gauge industry awareness and satisfaction using a process that enables us to check in with all registered participants.

Funding Mechanism and Limited Antitrust Exemption

Each state's law requires that MRC set a per unit consumer fee to fund its mattress collection, transportation, recycling, education, outreach, and administrative activities. The fee also funds each program's start-up costs and provides a reserve that allows the program to operate in a prudent and responsible manner over the long term.

To enable MRC to develop and implement a program plan (including setting and collecting the fee to fund the program), the state laws provide limited antitrust immunity to MRC, manufacturers, retailers, and other sellers for activities taken pursuant to the law. The public policy supporting the antitrust immunity provision is that the state's interests are served if mattress manufacturers can manage and fund an environmentally sound and cost-effective recycling program.

The fee is collected when residents of a state with an MRC recycling program buy a mattress or foundation. Although the per unit fee will vary from program to program (reflecting the different operational costs MRC incurs in each state), the fee is the same for all product types (both mattress or box spring) and sizes sold to residents in each program state.

Retailers (including traditional brick-and-mortar retailers as well as online and other types of retailers) must collect the fee on every mattress and foundation that they sell to consumers in a program state (regardless of where the seller may be based). Similarly, companies that sell these products to large-volume purchasers for delivery in a program state (including hotels, universities, military bases, furniture rental companies, etc.) must collect the fee. The fee must be itemized on the invoice or sales receipt provided to the purchaser. Each month, the retailers and other sellers remit the fees that they collect to MRC.

Beyond Recycling

The experience the industry has gained from establishing MRC has raised awareness across the mattress supply chain about the challenges existing products pose when they are discarded at the end of their useful life. It has also stimulated increased collaboration among component suppliers and mattress manufacturers and retailers as the industry seeks to improve the sustainability of its products and manufacturing methods.

In addition to its recycling activities, MRC also devotes resources to

- help mattress manufacturers reduce their environmental footprint,
- encourage the mattress supply chain to consider end-of-life challenges when designing new products and materials, and
- inform consumers how to extend the useful life of their mattresses through proper maintenance.

Sleep Products Sustainability Program

MRC developed the Sleep Product Sustainability Program (SP2), an industry certification program that helps manufacturers reduce manufacturing waste and increase operational efficiencies. Through SP2, companies establish or enhance their environmental management system to identify steps they can take to reduce environmental impacts, develop best practices and define metrics to track progress. SP2 can also serve as a foundation for pursuing an ISO 14001 certification, the international standard for effective environmental management systems.

Participants have experienced measurable reductions in manufacturing waste and energy. Training, program resources and certification are all offered at no cost to eligible mattress manufacturers.

MRC Industry Workgroups

MRC formed two workgroups that facilitate information sharing across the mattress industry value chain, including mattress manufacturers, component suppliers, and recyclers, to improve the recyclability of mattress materials.

MRC Industry Workgroup on Circular Design

MRC assembled the MRC Industry Workgroup on Circular Design to raise awareness among mattress component suppliers and mattress manufacturers about end-of-life recycling challenges when they design new components and mattresses. The workgroup initially focused on adhesives and continues to engage with members of the value chain for the purpose of:

- Discussing whether current materials and assembly techniques will pose recycling challenges when mattresses reach the end of life
- Informing suppliers, designers, and manufacturers about recycler deconstruction challenges and discussing how product design affects post-consumer materials management
- Exploring methods to improve the circularity of mattresses by reducing manufacturing waste and encouraging design and manufacturing practices that take into account the recycling value of post-consumer mattress materials, while balancing product function and durability

MRC Industry Workgroup on New Materials

The MRC Industry Workgroup on New Materials is composed of representatives from foam, textile, and spring manufacturers. The work group's objectives are to identify end-of-life challenges posed by current products and materials used in mattresses, advise MRC on anticipated long-term changes in product mix, and raise industry awareness about how design and marketing decisions can affect recycling.

What's Next

MRC fosters innovation, cooperation, and communication among suppliers, manufacturers, recyclers, and end-market commodity buyers around the world by working with industry experts and sharing information on efforts to develop a more circular economy for polyurethane foam and other mattress components. The organization continues to work to improve consumer awareness and access, used mattress collection and transportation logistics, and recycling efficiency and end-markets. The industry will also continue to work with states that are interested in enacting consistent and effective mattress recycling laws.

Part Two: Putting to Rest an Issue of Mattress Management

Thomas Metzner

More than 50,000 mattresses are discarded each day in the United States yet relatively few of them are recycled. Instead, most go straight to landfills. Due to the way mattresses are constructed, it's usually more expensive to recycle mattresses than dispose of them. The national average cost to dispose of a mattress is between $40 and $120. This high cost is an impediment to proper disposal and, as such, encourages illegal dumping of mattresses countrywide. Mattresses are expensive to manage because of their size and construction. They can be difficult to recycle. For example, mattress springs can get tangled in processing equipment. The cost of recycling a mattress is also high due to the amount of labor needed to deconstruct a mattress, and the revenues for various components are not sufficient to cover these costs. Mattresses also take up a large volume of space in a landfill relative to their weight.

Municipalities have typically been responsible for managing discarded mattresses. They have several options, such as collecting discarded mattresses curbside with other bulky waste and then disposing of them or having residents discard them at a transfer station for disposal. Municipalities may choose to use their general funds to finance mattress disposal or recycling by charging residents a fee to cover the cost. Notably absent in the past, when it came to providing a solution to managing discarded mattresses, were the mattress producers. While they design, engineer, market, distribute, and sell mattresses, until recently they have had no responsibility for managing their products at end-of-life. Involvement in the process, including product design (i.e., creating a more sustainable mattress) was needed.

Extended Producer Responsibility in Connecticut—Some Background

In my role as an analyst with the Connecticut Department of Energy and Environmental Protection (CT DEEP), I am responsible for evaluating strategies for sustainable materials management. Extended producer responsibility (EPR) has been a strategy we have used before with paint and electronics; mattresses seemed to be a good fit as well. Our first taste of EPR came in the mid-1990s when the battery industry developed a product take-back and recycling program, set up primarily through retail establishment drop-off. Connecticut has a law on the books going back to 1995 requiring that rechargeable nickel-cadmium batteries be feasibly removable from products and that municipalities had to provide for the recycling of these batteries "when service became available." New Jersey, Vermont, and Minnesota had requirements for recycling these batteries. In response, the battery manufacturers established a product take-back program at their expense, which allowed Connecticut municipalities to recycle nickel-cadmium batteries without incurring expenses. This battery law laid the foundation for Connecticut's advocacy for EPR.

Connecticut, like many states, disposed of its solid waste in municipal landfills from the 1940s through the 1980s. Almost every town in Connecticut had a solid waste landfill and many had more than one. Starting in the 1960s, when many landfills began to close, an alternative strategy for managing waste needed to be devised. Throughout the 1970s and 1980s, various plans and commissions were formed for the purpose of managing Connecticut's solid waste. The chosen strategy was source reduction followed by composting (when possible), mandatory recycling, and waste-to-energy (i.e., incineration with energy capture to generate electricity) for the materials that couldn't be recycled or composted.

One of the Connecticut waste-to-energy plants began operation in the South Meadows section of Hartford in the mid-1980s. The plant utilized refuse-derived fuel (RDF) technology, which processed the incoming waste into a fuel pellet before being fed into the incinerator. Unlike "mass burn plants" that can burn pretty much anything, RDF had certain waste streams that were not processable. These streams included scrap metal, bulky wastes such as furniture, and mattresses that had to be separated and shipped to a bulky waste landfill for proper waste management.

The City of Hartford operated a solid waste landfill in the North Meadows section of the city. It opened in 1940 and in the 1980s was leased by the city to the Connecticut Resources Recovery Authority (CRRA). CRRA was the quasi-state agency established by statute to plan for managing the state's solid waste. CRRA also owned the RDF plant in South Meadows. The Hartford landfill stopped accepting municipal solid waste in 2008. Until 2011, the landfill accepted bypass materials (wastes that were not able to be processed, such as mattresses) from the South Meadows RDF incinerator at no additional cost to the participating municipalities. But when the landfill stopped accepting bypass materials in 2011, CRRA needed a disposal outlet. Towns that had not previously paid for managing mattresses were suddenly looking at a significant expense. The City of Hartford alone projected that they generated 18,000 mattresses per year and at approximately $10 per mattress they were looking at a bill of close to $200,000 per year just to manage mattresses. Many other towns had similar issues and were facing high costs to divert mattresses to alternate disposal locations. Some towns could pass that cost on to residents who dropped off their mattresses at a transfer station, but not towns that had provided curbside collection, as Hartford had.

By this time, Connecticut had experience with EPR programs. EPR for paint passed in the state in 2011, preceded by Connecticut's first EPR program for electronic waste (e-waste) in 2007. For both programs, municipalities were able to accept electronics and paint at their transfer stations without having to pay for collection. Municipal leaders were beginning to see the benefits of EPR, and CT DEEP had adopted EPR as a key strategy in its solid waste management plan. Public works director for the City of Hartford, Marilyn Cruz Aponte, immediately recognized EPR as a solution for managing mattresses even though no other state (or Canadian province) had passed EPR legislation for mattresses.

With funding provided by the City of Hartford, the Product Stewardship Institute (PSI), a nonprofit organization that works with various entities including state and local governments advocating for EPR, was retained to begin a process of engagement with stakeholders with the idea of passing the nation's first EPR law for mattresses. Stakeholders included municipalities, the trade association representing mattress manufacturers, and CT DEEP. PSI had been instrumental in getting the paint manufacturers to buy in to EPR for paint and had been working with many state and local governments on EPR initiatives for a variety of waste streams including e-waste, thermostats, and pharmaceutical drugs. PSI's process is to begin with an identification of the problem to be solved, make a list of impacted stakeholders, and then to convene a working group geared toward finding acceptable solutions. This stakeholder working group then works on the outline of an agreement for the basic elements of a program that can then be turned into a draft bill.

The intent of CT DEEP and the City of Hartford was to undertake a process to pass EPR legislation that was similar to the paint process the state had just gone through. However, the mattress industry, represented by the International Sleep Products Association (ISPA), clearly had different interests. At the first meeting in Hartford in 2011, ISPA representatives would not entertain thoughts of EPR for mattresses. From their perspective, mattress EPR had never been done before and it appeared they believed they lacked the expertise to manage a recycling program. They mentioned the expense associated with transporting and processing of discarded mattresses.[4] ISPA seemed to be saying that operating a mattress recycling system would be too

[4] International Sleep Products Association. Member Alert Regarding Mattress Recycling Legislation. Sleep Products, 2012. https://sleepproducts.org/2012/05/ispa-advocacy-alert-member-alert-regarding-mattress-recycling-legislation (accessed 2022).

expensive and therefore they wouldn't do it. They considered it a municipal issue and had very little to offer the towns around the table. What is frustrating for municipalities when this happens, whether it be in an instance like this or in managing waste in general, is that they have no option but to address the issue: materials discarded by residents must still be collected. While municipalities couldn't simply opt out of managing this material, they did have leverage: passing EPR legislation for mattresses.

The mattress EPR bill was first introduced in Connecticut in 2012. Elements of this bill included a plan for financing the collection and recycling of mattresses, building awareness, and educating consumers to promote program participation, and evaluating progress. Although the bill didn't pass, there was significant support from municipalities and key legislators. In the Connecticut legislature, even numbered years are short sessions and so passing certain bills can be a challenge. Long session years, like 2013, are usually more favorable for a bill such as EPR for mattresses. It became evident to ISPA that there was a very good chance an EPR bill would pass in 2013 given the legislators who were supporting it. It appeared that the votes to pass the bill were there. The same sponsors for the paint EPR bill were also supporting the mattress initiative. The similarities between the bills were evident, and the paint bill had passed unanimously. The main request from ISPA was that they be allowed to cover the program costs by assessing a fee at point-of-sale, charged to the customer. Since the paint industry had a similar provision, called an eco-fee, the provision was included.

Mattress Stewardship Law

The bill did indeed pass into law in May 2013, making it the first mattress EPR law (Public Act 13–42) in the United States. With this legislation, mattress manufacturers are responsible for properly managing mattresses statewide in Connecticut.

> Such mattress stewardship program shall, to the extent it is technologically feasible and economically practical: (1) Minimize public sector involvement in the management of discarded mattresses; (2) provide for free, convenient

> and accessible state-wide opportunities for the receipt of discarded mattresses from any person in the state with a discarded mattress that was discarded in the state, including, but not limited to, participating covered entities that accumulate and segregate a minimum of fifty discarded mattresses for collection at one time and municipal transfer stations that discard a minimum of thirty mattresses at one time; (3) provide for free collection of discarded mattresses from municipal transfer stations that accumulate and segregate fewer than thirty mattresses, provided the transfer stations require such collection due to space or permit requirements; (4) provide for council-financed end-of-life management for discarded mattresses collected pursuant to subdivisions (2) and (3) of this subsection; (5) provide suitable storage containers at, or make other mutually agreeable storage and transport arrangements for, permitted municipal transfer stations for segregated, discarded mattresses, at no cost to such municipality, provided such municipal transfer station makes space available for such purpose and imposes no fee for placement of such storage container on the municipal transfer station's premises; (6) include a mattress stewardship fee that is sufficient to Substitute House Bill No. 6437 Public Act No. 13–42 6 of 12 cover the costs of operating and administering the program; and (7) establish a financial incentive that provides for the payment of a monetary sum, established by the council, to any consumer who recycles a mattress in accordance with the requirements of the mattress stewardship program.[5]

The driving force for passage was the City of Hartford and their public works director, Marilyn Cruz Aponte. She organized various stakeholders who supported the bill and secured the votes of key legislators. Pat Widlitz, the Connecticut state representative from the shoreline towns of Branford and Guilford, was the key legislative proponent. She also led the passage of Connecticut's e-waste and paint EPR laws and was a respected and committed legislator. Shortly after Connecticut passed its law, Rhode Island and California followed suit.

There are a couple of take-aways from the Connecticut mattress law process. First and foremost is to find the right legislative advocate. Second is to get

[5] State of Connecticut. Substitute House Bill No. 6437, Public Act No. 13–42. An Act Concerning a Mattress Stewardship Program (October 1, 2013). https://www.cga.ct.gov/2013/act/pa/pdf/2013PA-00042-R00HB-06437-PA.pdf.

key municipal leaders to testify. The City of Hartford's mayor, Pedro Segarra, testified before the environment committee telling the story of how mattress EPR would benefit his city. Third: good data is also important. We were able to calculate that implementing EPR for mattresses would save municipalities $1.2 million annually, as well as create jobs while diverting recyclable materials from the municipal solid waste stream.

ISPA was now fully on board with implementing the mattress EPR program. CT DEEP coordinated with ISPA on several stakeholder meetings to start the planning process. Because this was a first-in-the-nation law, we didn't have a template for getting started. We relied on the similarities with the paint program that had kicked off on July 1, 2013. Both programs had a fee at retail. The eco-fee was assessed by the stewardship organization, the Mattress Recycling Council (designated by ISPA to manage the state program), to cover costs of operating the program. Both programs concerned products that previously had little to low recycling rates yet were big issues with which to contend. Both programs had to be introduced to the public in terms of building awareness, knowledge, and specifics to engage in the program. But unlike the paint program, which did not have a collection infrastructure, in the mattress program, mattresses had already been collected at the curb and at transfer stations and so this aspect was in place and would not change. There was no need to develop a new collection method, simply to recycle what was collected by the generators that included residents, hotels, schools, hospitals, and volume reduction facilities.

Municipalities were not mandated to participate; this was key to getting support for the law. It would seem intuitive that a municipality would want a different entity to cover the expense of managing discarded mattresses, but some municipalities had reservations. Some towns didn't have any expense associated with disposal because they assessed a fee to residents at the transfer station. Other towns expressed concerns with the requirements for participating that were imposed by the stewardship organization. This included requiring that mattresses be stacked in a certain way within the roll-off container. Transfer station attendants don't typically assist residents in unloading materials from vehicles or loading them into roll-offs. There were also concerns about potential impacts of mattresses infested by bed bugs. Some towns also lacked the space at their transfer stations for containers to store the mattresses. In addition to these concerns there were towns that mentioned they would like to witness the program in operation before signing up.

A key component of the plan was to identify the facilities that would process mattresses. Before the bill passed, there was a mattress recycler in Bridgeport, Connecticut—Park City Green—that had been processing mattresses. Park City Green had an important social mission; they paid employees a living wage and gave jobs to men and women who had difficulty securing employment. But they had to compete with disposal of mattresses and most towns in the area would incur a cost increase to pay for recycling. A facility had opened in Bloomfield, Connecticut, in anticipation of the EPR law passing. They were experienced in mattress recycling, having set up facilities in Canada and France. The partners in this business had a parting of ways and the remaining owner moved to work at the facility to East Hartford. These two facilities were selected by the ISPA stewardship organization to process all of the mattresses in Connecticut. To provide an idea of the magnitude of the program, in 2022 the number of collected mattresses was 205,000.

The stewardship organization, the Mattress Recycling Council, submitted their plan in 2014 for recycling mattresses in Connecticut. The CT DEEP approved the plan on December 31, 2014. The program officially kicked off in the spring of 2015 with an event at the Hartford transfer station, a site that was situated next to the closed landfill that started this whole process. The program had consistent growth through its first years, both in the number of towns participating and mattresses collected. MRC implemented an awareness campaign that included public service announcements and billboards, in addition to posts on social media. Within five years, 136 of Connecticut's 169 municipalities were participating in the collection program. Most mattresses that were collected came from municipal transfer stations in the state. Before the program started, municipalities typically charged $10–$30 to a resident who was dropping off a mattress. Once the program started, those fees went away because the cost of managing mattresses had been transferred from the municipality to the stewardship organization. This eliminated a financial barrier that had possibly impeded resident participation to an extent in the past.

Connecticut's leadership in implementing EPR for mattresses laid the groundwork for other states. Rhode Island and California passed similar legislation shortly afterward and Oregon became the fourth state, establishing EPR for mattresses in 2022. Setting goals for the program was challenging because there was insufficient information on the number or sources of generated mattresses. In addition to collecting mattresses from residents who had delivered them to transfer stations, we knew that mattresses were also generated from residential schools, hotels, hospitals, and prisons, but it was while the program was running that CT DEEP learned more about mattress generation sources. In addition to households, many also come in through retailers that accept or "take back" a mattress when they deliver the new one. Hotels generate many mattresses. These are of high quality, and many are sold to the public and brokers for resale to the public. Universities and boarding schools take advantage of this program when they do mass changeovers in dormitories, but they don't have space to store mattresses that are intermittently removed. Mattresses from hospitals are specialized and very expensive and so they are usually sold privately. Prison mattresses don't contain any metal or wood for safety reasons and have not been recycled through this program. Other mattresses are too contaminated to be recycled and they are disposed of as waste.

Connecticut conducted an evaluation of the mattress EPR program in 2022.[6] The evaluation included establishing collection and recycling goals, an evaluation of the eco-fee, and recommendations to improve the program. While the program overall has been successful, there are areas that can be improved. The single best way to improve collection is to offer curbside collection. Curbside collection of mattresses is the most convenient approach and can increase program participation. Municipalities that collect mattresses curbside for recycling have higher collection rates than those that do not offer curbside collection or only offer it for disposal.

There are barriers for residents to deliver mattresses to collection points. Many city residents for example lack transportation to bring a mattress to a transfer station. Others have difficulty dealing with such a bulky item.

[6] Connecticut Department of Energy and Environmental Protection, Evaluation of Connecticut's Mattress Stewardship Program, January 2023. https://portal.ct.gov/-/media/DEEP/reduce_reuse_recycle/mattress/Mattress_EPR_Report_012523.pdf.

A mattress may not fit in their car and tying it to the roof may not be a comfortable option.

They also didn't want to change a collection method that existed before the start of the program. If residents were accustomed to putting mattresses curbside, municipalities were not going to change that. However, mattresses that are placed outdoors for collection are also subjected to rain. Processing wet mattresses is a challenge for recyclers. They need to be dried before processing and that requires storage space, which is not available for one of the two recyclers.

As mentioned, the mattress EPR program, similar to the paint EPR program, is financed by an environmental handling fee ("eco-fee") assessed on mattress purchases at retail sites. This eco-fee is charged to fund mattress recovery and recycling. The paint and mattress manufacturers insisted on this method of cost externalization as a condition of their support. Initially, some consumers believed this was a tax that was remitted to the state. Over time, calls about the fee from the public have stopped, but the debate continues about whether the manufacturers should assume the costs as part of putting their product on market or pass it on to consumers. There is also a concern about fairness of the eco-fee assessment to consumers because it is the same cost regardless of income level or the cost or quality of a mattress (e.g., twin discount sets are assessed the same fee as luxury king sets). It is apparent this fee is likely more of a burden on lower-income families. According to the Mattress Recycling Council, it costs the same amount to recycle different types of mattresses and often, lower-priced and lower-quality mattresses come into the system earlier than more expensive and durable mattresses (personal communication with Alison Keane, April 8, 2025).

Despite challenges, the Connecticut mattress EPR program has been a resounding success. We know that mattresses do not belong in landfills; they are light and bulky and take up space at landfill sites that have limited capacity. Mattresses also don't belong in waste-to-energy plants. Overall, they're a difficult discard to manage. Mattresses can be recycled, but it is labor intensive. Also, mattress springs get caught in equipment. A reason that there aren't many companies across the nation that recycle mattresses is because landfill disposal tends to be a more economical option in the short run. The best way to make mattress recycling a competitive option is to implement EPR. In 2022, Connecticut recycled more than 200,000 mattresses. Before EPR, only a small fraction of that number was recycled. EPR for mattresses has created or sustained tens of jobs and saved municipalities more than $12 million in disposal costs in Connecticut.

It is surprising that more states haven't adopted mattress EPR. From the CT DEEP perspective, this program has been a solution to the high cost of collecting and recycling mattresses that saves municipalities money. It would seem intuitive that states and municipalities would want to increase recycling of mattresses and avoid the expenses of disposal and recycling. A few states have called CT DEEP to inquire about our program. Massachusetts and Maryland have introduced legislation, but they haven't passed EPR for mattresses yet. Oregon is one recent state that has successfully passed it.

One argument against EPR that we have heard in other states is that residents would simply cross the border to purchase the mattress and avoid the fee. We tested this theory by calling mattress retailers on both sides of the Connecticut border: Massachusetts and New York. We found there was no change in purchasing habits related to EPR. The stores closest to the New York and Massachusetts borders didn't see a rise in business from Connecticut customers. Connecticut retailers likewise didn't report a drop in sales because of the law.

All the states that have mattress EPR had previously implemented paint EPR; that's not a coincidence. The paint EPR program has been very successful in Connecticut. It allowed residents to get rid of their paint conveniently, through a network of retailers. The success of the EPR for paint program was something we pointed to as a reason to use the same strategy for mattresses. Connecticut also had an existing electronic waste EPR program going back to 2009. Municipalities were required to provide e-waste collection for their residents and EPR allowed them to meet this requirement without incurring expenses. Continuing on this path, in 2022, Connecticut was the first state to pass EPR for propane gas cylinders. This growing familiarity with EPR programs has led to municipal support for EPR as a solution for other problematic wastes including tires, household hazardous waste, and packaging. The support for EPR extends beyond familiarity and building on prior successes. A reason that CT DEEP has been a strong proponent of this approach is because it will increase the diversion of waste from landfills and incinerators while saving municipalities money. There is no landfill space available in Connecticut and as the existing waste-to-energy plants age, we are faced with aggressive diversion and recycling programs or citing another facility. Shifting responsibility from municipalities to producers through EPR can ease the burden of material collection and recycling.

5

Designing Behavior Change Programs to Reduce Waste

Jennifer J. Tabanico
Action Research

The root cause of nearly every environmental problem is humans acting in specific ways. That is, environmental problems are behavior problems. Consequently, knowing how to effectively change behaviors that impact the environment is central to solving many of the environmental problems we face locally, nationally, and globally (Williamson et al., 2018). Over the past two decades, I have worked with dozens of agencies to address various environmental problems, including water quality, air pollution, climate change, waste reduction, human–wildlife conflict, biodiversity, and more. My clients are widely diverse in terms of the environmental issues they are trying to address, the audiences they are trying to reach, and the behaviors they are trying to change. They have represented every level of government, from federal agencies to local townships, universities, community organizations, and private businesses. They have worked to develop programs for a range of audiences, including youth, college students, homeowners, renters, business owners, farmers, and building operators. They have strived to engage people in simple repetitive behaviors such as picking up dog waste or taking reusable bags to the grocery store and complex one-time behaviors such as installing energy-efficient windows or replacing turf with drought-tolerant plants. Despite this diversity, they have all come to my firm with the same, often desperate, pleas:

- We need to move the needle!
- How do we get people to take action?
- What is the key message we need to tell people to get them to change their behavior?

Underneath these pleas has been a shared recognition that their work to engage the public through websites, fliers, workshops, advertisements, and other information-intensive approaches was often successful in generating awareness or knowledge about the environmental problem and its associated behavioral solutions but rarely translated into the behavior changes needed to make an impact.

I am not an environmental scientist. I did not study environmental law or policy. So why are so many agencies increasingly flocking to behavioral scientists like me for help? Because there is a growing recognition that we cannot simply engineer, educate, or regulate our way out of pressing environmental problems without effectively addressing behavior. Tackling these pressing problems requires the development of solutions that are grounded in the way individuals think, feel, and act.

In this chapter, I present a framework for developing behavior change programs that is broadly applicable across diverse environmental domains. Programs developed using this framework have produced measurable results across a wide range of environmental behaviors across the globe (McKenzie-Mohr, 2011). This chapter focuses specifically on community behavior change approaches that government and municipalities, nonprofits (including producer responsibility organizations), businesses, solid waste educators, and others can employ to improve municipal solid waste (MSW) reduction and material resource recapture (e.g., recycling and composting) programs. It will provide an actionable and iterative process for going beyond raising awareness and knowledge to achieve measurable changes in behavior. The chapter begins by demonstrating the importance of behavior and why awareness and knowledge are rarely sufficient to "move the needle." I will then present an overview of a scientifically grounded framework for developing programs to change behavior. Following this overview, I'll demonstrate step-by-step how to apply this framework through illustrative examples and hands-on practice.

Principle #1: Behavior Matters

When it comes to individuals working on solving environmental problems, the *initial* tendency is rarely to recognize the issue as a behavioral problem. Engineers and scientists approach environmental problems by developing technical solutions such as programmable thermostats, solar panels, and electric vehicles. Planners approach environmental problems with the *Field of Dreams* mindset ("If you build it, they will come"). They offer infrastructure solutions like bike lanes, household hazardous waste (HHW) collection facilities, and curbside food scrap collection programs. Policymakers approach environmental problems by creating and implementing policies—legislation and guidelines—that address contributing issues such as plastic debris (e.g., plastic bag bans) and an overabundance of product packaging (e.g., a law to reduce the amount of allowable packaging). Regulatory bodies drive the development and enforcement of state and local laws and regulations surrounding environmental protection, such as fines for illegal dumping or irrigation runoff. Although technological advancements, infrastructure, policy, and law play a crucial role in addressing environmental problems, these approaches cannot ignore the relevance of behavior change.

Behavior is embedded within each of these solutions to environmental problems:

1. Technology—New technologies and products will not have their intended effects if people are unwilling to use them regularly (e.g., reusable straws) or if they misuse them (e.g., businesses using compostable containers where adequate composting infrastructure does not exist).
2. Infrastructure—Adequate infrastructure is essential to support behavior change (e.g., curbside food waste collection). Yet if the public doesn't engage with and correctly use the infrastructure (e.g., removing contaminants such as plastic packaging), the needle will not move.
3. Policy—Policy changes take considerable time to build and implement, often requiring a great deal of public support. Getting people engaged in the behaviors voluntarily (e.g., using reusable bags at the grocery store) can generate public support and produce meaningful progress until policies (e.g., plastic bag bans) are in place.

4. Law—Enforcing laws is an effective way to garner compliance, but it also takes considerable time and resources. For many behaviors, catching people "in the act" (e.g., littering or illegally dumping) is difficult, making enforcement cumbersome. Other behaviors (e.g., flushing unwanted pharmaceuticals) are essentially impossible to enforce without the help of high-tech monitoring devices.

Because environmental problems are behavior problems, the solutions must also address human behavior (Schultz, 2011). Therefore, a critical first step to effectively developing programs that address resource consumption and waste is acknowledging that behavior matters. These problems are behavior problems, and human behavior plays a role throughout the product chain, from purchase to disposal or recycling.

Principle #2: The Behavior Matters

Now that we have acknowledged that behavior matters, we need to go further to recognize that *the particular* behavior matters. Within the environmental domain of waste reduction and waste management, there are hundreds of behaviors that could potentially be targeted with a behavior change program. These include behaviors as diverse as recycling used mattresses, disposing of leftover paint, recycling used motor oil, reusing bags for groceries, using durable utensils for takeout food, placing food waste in a curbside collection container, backyard composting, placing clean recyclable beverage containers in the appropriate recycling bin, and buying products with minimal packaging, to name just a few (McKenzie-Mohr et al., 2011; see also www.cbsm.com). Each of these behaviors varies with respect to its impact on waste outcomes.

Additionally, the barriers that inhibit, and the drivers that encourage, behavior often vary at the level of individual actions. For example, the barriers to properly recycling a used mattress (e.g., inconvenience or lack of knowledge about disposal options) are likely very different than the barriers to using reusable utensils while dining out (e.g., forgetting or worrying about the perceptions of others). Perceived benefits, though to a lesser extent, may also vary at the level of individual actions. For example, an individual might be driven primarily by the financial benefits of participating in reuse behaviors (e.g., secondhand shopping). In contrast, that same individual may be driven by the social benefits of participating in a community repair and reuse workshop or the safety benefits of taking used lithium batteries to a collection point. Whether these barriers and benefits are actual or perceived, internal to the individual (e.g., beliefs) or external to the individual (e.g., infrastructure), they can increase or decrease the likelihood of individual behavior change.

The specific behavior that will be the focus of your program matters because not all behaviors are created equal in terms of their impact on resource consumption and waste management outcomes. Also, the barriers and benefits vary widely across behaviors and contexts. Consider a visit to your local coffee shop, as I did. I took a quick inventory of the desirable behaviors one might engage in within that context to address the environmental problems caused by single-use plastics. They included bringing in a reusable mug for hot coffee or a cup for cold or blended beverages, refusing a plastic straw, bringing reusable utensils, buying food without plastic packaging, and placing recyclable packaging (without food or liquid residue) in the correct bin. Try this the next time you are at a coffee shop, sports venue, or cafeteria. As waste-savvy readers, you'll likely come up with needed changes that could be the focus of behavior change programs to reduce waste. Each of the six behaviors I noted has different barriers and waste reduction potential. Admittedly, while my motivation for personally engaging in any of these behaviors is a strong desire to reduce waste and protect the environment, the barriers I face to each of these vary widely. Bringing in my own mug for the house coffee is widely accepted and relatively easy (if I can remember to grab my mug before leaving the house), but if I plan to treat myself to one of those icy blended coffees, my mug will not do, and it is going to be hard to say "no" to the straw.

Whether you are reading this chapter as a policymaker or regulator, municipal official, educator, brand owner, planner, marketer, or other stakeholder, you are on your way to understanding and developing a successful program by recognizing that behavior matters and the particular behavior matters more. This may seem like an obvious distinction. However, identifying the "best" behavior upon which to focus is strategic and not as simple as it may seem.

Information-Intensive Approaches to Behavior Change

Most efforts to promote environmentally responsible behaviors have traditionally utilized marketing campaigns and wide-reaching media efforts driven by an information deficit model. This model assumes that the primary reason that people don't engage in desired behaviors is that they don't know they should (or should not) engage in a behavior or what specifically to do (Burgess, Harrison, and Fillius, 1998; Simis et al., 2016). This leads to an exclusive focus on informing people about the issue and raising awareness about its importance. While knowledge and awareness are important targets, people engage in behaviors for various reasons, and simply knowing what to do is not a motivation for action (McKenzie-Mohr, 2011; Schultz, 2002). Although these campaigns are well intended, a growing body of scientific research demonstrates that providing information rarely results in individuals adopting or modifying behavior, even in the most highly visible, appealing, and well-funded campaigns (Christiano and Neimand, 2017). For example,

- The widely recognized National Cancer Institute's "5 a Day" campaign successfully increased knowledge from a baseline of seven percent of US adults who knew they should eat at least five servings of fruit and vegetables per day to 20 percent who knew six years into the campaign. Sadly, actual fruit and vegetable consumption remained static at 11 percent over an eight-year period (Casagrande et al., 2007).
- Well-known anti-drug and anti-smoking campaigns like the Drug Abuse Resistance Education (D.A.R.E) program have not only been ineffective at reducing adolescent substance use, but some even led to increased use. These campaigns piqued curiosity and led teens to believe that substance use behaviors were more common among their peers than they were. Lack of knowledge was not the problem. In fact, teens were more influenced by social norms and a desire for rebellion (Hornik et al., 2008). This mistake of not understanding the audience's needs and motivations cost taxpayers millions of dollars.
- Over a ten-year period, California spent millions of dollars on an advertising campaign, "Flex Your Power," to promote energy conservation

actions by residents. The program relied on informational mass media messages and did not target barriers to specific home energy conservation actions. As a result, the campaign failed to produce substantial changes in behavior (Opinion Dynamics Corporation, 2010), resulting in a negligible return on investment (ROI) for their media spend.

Waste reduction and management behaviors are often more difficult, costly, or time-consuming than the consumption-driven alternatives, and simply knowing what to do or recognizing their importance is rarely enough to spark action. Lack of knowledge and unsupportive behaviors represent only two barriers to behavior change. There are many other external (e.g., lack of infrastructure) and internal (e.g., fear of social disapproval) barriers that prevent the adoption of the desired behaviors.

Overview of Community-Based Social Marketing

At this point in the chapter, you may be feeling a little discouraged, asking yourself, "if these programs can't change behavior despite many thousands of dollars in advertising spend, how am I supposed to do it? Or, how am I supposed to do it with a handful of staff and a shoestring budget?" Fortunately, there are viable alternatives that can engage and move your audiences toward actions that are critical to your program goals while, at the same time, maximizing valuable and limited resources.

Community-based social marketing (CBSM) is one such approach that has emerged as an effective alternative to traditional information-intensive campaigns (McKenzie-Mohr, 2000; McKenzie-Mohr et al., 2011; Schultz and Tabanico, 2008; Tabanico and Schultz, 2018). CBSM is not a media tactic or messaging strategy. It is an empirically derived and data-driven framework for developing behavior change programs that marries knowledge from the field of social marketing with a variety of behavior change tools drawn from the social and behavioral sciences. CBSM involves five steps:

1. Strategically selecting which behavior(s) to target;
2. Identifying the barriers and benefits to the selected behavior(s);

3. Developing strategies that reduce the barriers to the behavior(s) to be promoted while simultaneously enhancing the benefits;
4. Piloting the strategies and determining which are most cost-effective; and
5. Broadly implementing the most cost-effective strategies and conducting an ongoing evaluation (McKenzie-Mohr, 2011).

CBSM has been demonstrated time and again to produce substantial and measurable changes in behavior, and the approach is well suited to address the specific needs of those working to reduce and manage waste. Additionally, because the programs developed using this approach are initially piloted on a small scale, the program can be refined as needed until there is documented evidence to show that the program works before it is scaled up. This research-based approach can prevent spending time and money on ineffective programs or placing resources entirely in advertising campaigns or a social media presence that are unlikely to lead to measurable behavior change by themselves. Following is an overview of each step.

Step 1. Select Behaviors

A successful CBSM campaign must both (1) target behaviors that are linked to the desired outcome or goal and (2) target specific behaviors rather than make broad appeals (e.g., reduce waste). Even within a single sector (e.g., organic waste), there are numerous individual behaviors that residents or other audiences could engage in to achieve the desired outcome such as reducing food waste by using leftovers, backyard composting, placing food scraps in curbside collection bins, and proper food storage.

Behavior selection begins with the development of an exhaustive list of target behaviors that are linked to a particular outcome (e.g., organic waste in the landfill). It then proceeds with a careful analysis of the environmental impact of these behaviors, the existing level of engagement among members of the target population (referred to as penetration), the willingness of the audience to engage in the behavior (probability), and the relevance of the behavior to the target audience (applicability). Various data sources are considered in this analysis, including case studies, subject matter expert interviews, technical data, and quantitative survey data. It is important to be strategic about behavior selection to avoid spending money and resources on the wrong behaviors or the right behaviors at the wrong time.

Ultimately, the behaviors that are selected for CBSM campaigns are those that have the best combination of high environmental impact (i.e., the behavior is meaningfully connected to the problem), low rates of penetration (i.e., there is room to move people), high probability of engagement (i.e., there is a readiness to change), and broad applicability (i.e., the behavior is relevant to many). Later in this chapter I present ways to gather the information you need to strategically select target behaviors using these factors.

Step 2. Identify Barriers and Benefits

There are hundreds of behaviors that share the common goal of reducing waste. However, the barriers that inhibit engagement vary at the level of the specific action. Therefore, once you have identified which behavior or small set of related behaviors you wish to focus on, the next step is to identify the factors that decrease the likelihood of behavior change (barriers) and the drivers that increase the likelihood of behavior change (benefits).

The CBSM approach recognizes that perceived barriers and benefits to engaging in environmental behaviors vary depending on the population, context, and behavior of interest. Additionally, multiple barriers can exist simultaneously for each behavior. For example, the factors that prevent an individual from bringing unwanted pesticides to a household hazardous waste (HHW) collection event (e.g., transport or safety concerns) are likely different than concerns about bringing in batteries (e.g., lack of knowledge about what types of batteries are accepted). Barriers can be either internal to the individual (e.g., lack of knowledge or perceived social disapproval) or external to the individual (e.g., structural elements of the program, such as collection facility distance and operating hours). They may also vary as a function of some audience characteristic (e.g., homeowners vs. renters). While it is possible, and all too common, to develop a list of things that we *think* stand in the way of individuals taking action, it is important to set aside personal speculation and base campaigns on input collected directly from the target populations. Indeed, the factors that the audience perceives to be beneficial about engaging in the behavior are often different from the reason for delivering the program. CBSM involves identifying barriers and benefits through carefully conducted research, including reviews of existing literature, observations, intercept interviews, focus groups, and surveys.

A fundamental element of this step in the CBSM process is to identify what differentiates people who are currently engaged in the target behavior (the doers) from those who *do not* engage in the target behavior (the non-doers). Therefore, barrier and benefit research efforts must include gathering data from both groups. For example, suppose the program's goal is to collect unwanted pesticides at an HHW collection facility. In that case, it is important to identify the barriers perceived by people who *do not* already use the facility. Surveys or interviews with individuals at the collection facility will provide useful information, but we cannot assume that the findings will generalize to the broader population. Indeed, the fact that they showed up to the facility when others did not suggests that they are inherently different. A more suitable approach in this case might be to interview individuals as they are leaving a lawn and garden center with pesticide purchases. This action would more likely result in a sample that includes those who properly dispose of unwanted pesticides at a collection site and those who use other disposal methods.

While it can be tempting to forgo audience research, especially when time or budget concerns are of the essence, do NOT skip this important step! Uncovering barriers and benefits is a hallmark feature of the CBSM approach and an essential step in creating an effective outreach campaign or improving an existing program. Without a good understanding of what inhibits and encourages the target audience to engage in the target behavior, strategies will be based on (possibly faulty) assumptions and are unlikely to be effective. Understanding the underlying barriers and benefits to specific actions and their relative importance enhances the likelihood of developing programs that achieve behavior change. It is important to understand benefits. Benefits may include social approval, cost savings, keeping others safe, and other real or perceived positive outcomes of engaging in the behavior. Often, the audience's reason for, or primary perceived benefit of, engaging in the desired behavior (e.g., preventing harm to pets) differs from the reason for delivering the program (e.g., environmental protection). Therefore, program planners should be prepared to exercise humility and to design programs that speak to the benefits perceived by the audience.

Step 3. Develop Strategies

Once sufficient data is gathered on the barriers and benefits to the target behaviors, the next step is to select the behavior change tool(s) that address them and incorporate these tools into interventions (including program messages) that will have success in overcoming these barriers, enhancing motivation, and ultimately changing behavior.

The first question that needs to be answered is whether the reasons for not participating in the program are internal or external (or both) to the individual. External barriers are aspects of the program that decrease the likelihood that an individual will participate, sometimes called "structural barriers." For example, collection facility hours of operation, travel distance to collection sites, the difficulty of the behavior (e.g., requirements to wrap used mattresses in plastic), and complex collection methods are structural elements of the program that may represent barriers to participation. Internal barriers reside at the individual level, including psychological, emotional, social, and motivational barriers. Examples of internal barriers include unfavorable attitudes or misconceptions about the behavior, lack of knowledge about how to adopt the behavior, believing the behavior is unimportant, attentional capacity, fear, established habit patterns, or a perception that few others use the program (normative beliefs).

One of the greatest strengths of CBSM as an alternative to information campaigns is that it draws heavily on behavioral science research literature, particularly social psychology, to identify tools for overcoming internal barriers. These tools may include providing normative information, using (verbal or written) commitments, promoting social diffusion, fostering habits, and using prompts or incentives. These and other behavior change tools and the context in which they can be most effectively applied to remove barriers or enhance motivation are addressed later in the chapter. When designing strategies, one should consider both barriers and motivation (McKenzie-Mohr and Schultz, 2014). If barriers are high, structural changes that increase the convenience of engaging in the behavior will likely be needed. Efforts

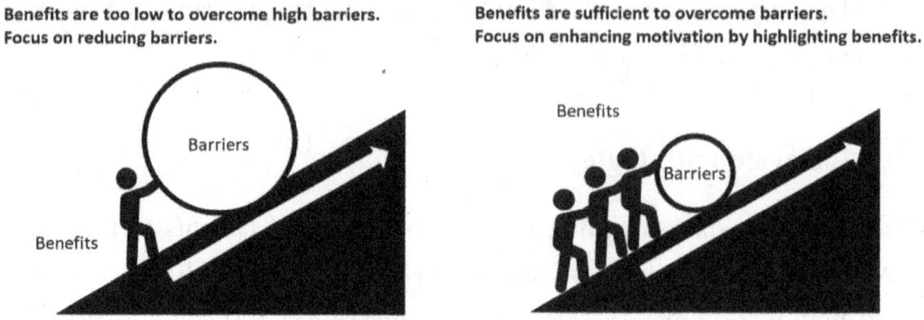

Figure 5.1 Addressing barriers and motivation in CBSM campaigns.

exclusively focused on motivation are unlikely to offset the deterring power of the barriers. Conversely, if barriers are low, program resources directed at enhancing motivation by highlighting benefits could be sufficient to drive behavior change (Schultz, 2014). See Figure 5.1 for a graphical representation of the relationship between barriers and benefits in developing CBSM strategies.

Step 4. Pilot Test

Once the behavior change program has been designed, the fourth step is to pilot the intervention strategy (or strategies) with a small segment of the audience using a control group, if possible, that does not receive the intervention. Including a control group provides a standard for comparison and helps to account for seasonal effects or other influences (e.g., news events). Pilot testing allows program planners to determine the efficacy of specific programs and highlights areas where changes need to be made before widespread implementation. Furthermore, testing several strategies against each other on a small scale is a useful way of identifying the most cost-effective approach to achieving behavior change. If the pilot program is unsuccessful, the strategy should be refined and then piloted again until it achieves demonstrated behavior change.

Step 5. Broad Implementation and Evaluation

When the pilot is successful at changing behavior, the most cost-effective strategy can be implemented more broadly across the community. Additionally, strategies that are developed and pilot-tested using the CBSM framework can be packaged into turnkey strategies or toolkits that can be applied more broadly to groups that share similar profiles of barriers and benefits to the target behaviors. Importantly, in scaling up the program, agencies should include all program elements that were included in the successful pilot. For example, if a successful pilot included in-person communication about the importance of recycling household batteries, we could not assume that we would get the same effect if the same information were delivered more passively through direct mail. Similarly, if the pilot included providing residents with a storage bin for storing and transporting household batteries, then we could not assume the same level of participation if this bin was not provided.

The takeaway message here is to only pilot test strategies that can be sustained on a larger scale. Community-wide implementation should be carefully evaluated by comparing baseline measures of behavior to behavior at several points following the intervention. If behavior change begins to wane, you may need to adjust the strategy and address new barriers (e.g., changing political climate or new materials in the market) or support motivation (e.g., with feedback) until new habits are formed that support the longevity of the behavior change.

These five CBSM steps can be applied to address an endless list of behavior change challenges faced by those working in municipalities or stewardship organizations, such as addressing product waste reduction and resource material recapture. Table 5.1 provides an illustrative example of how the steps might be applied to the goal of increasing the capture rate of household batteries.

Utility of Community-Based Social Marketing for Practitioners

Since its introduction in 1995 by Doug McKenzie-Mohr (McKenzie-Mohr et al., 1995; McKenzie-Mohr, 2011), the CBSM approach has become broadly accepted and is now being utilized by a wide variety of governmental and nongovernmental organizations (see www.cbsm.com and www.toolsofchange.

Table 5.1 Sample Application of the Five CBSM Steps

Program Goal: Increase the community capture rate of rechargeable household batteries.
Sector: Residential

CBSM Steps	*Activities*	*Program Actions*
Step 1: Select Behaviors	Review existing collection, waste characterizations, and other relevant program data to determine how many and what type of batteries could be captured instead of improperly disposed (impact). Determine through surveys or existing program reviews the extent to which various battery types are already collected (penetration) and the willingness among members of the target population to take rechargeable batteries to a collection site (e.g., retail location) compared to other available forms of collection (e.g., community collection events).	You have evidence that a significant number of rechargeable batteries are not collected and are either stored or placed in the trash. You conclude there is low utilization of collection sites, particularly for cell phone and laptop batteries, yet there is a willingness to use them. Therefore, you decide to focus your program on encouraging residents to deposit their unwanted, spent cell phone and laptop batteries at designated collection centers in the community.
Step 2: Identify Barriers and Benefits	Identify articles, case studies, technical reports, and other related work to identify what barriers and benefits have been identified by other rechargeable battery collection programs or related programs (literature review). Using the research literature as a foundation, develop and administer a survey of residents to identify current practices as well as barriers and benefits of taking cell phone and laptop batteries to a collection center.	Since the behavior has broad applicability (most households are likely to have batteries from cell phones or laptops), you decide to administer a brief survey by mail to a random sample of residents living in neighborhoods within proximity to collection points. Most people believe that taking rechargeable batteries to a collection center is the right thing to do, but they store their batteries at home in a drawer and forget to take them to a collection site when they leave home. They are also uncertain as to what types of batteries are accepted at these sites or if there is special packaging required.

Program Goal: Increase the community capture rate of rechargeable household batteries.
Sector: Residential

CBSM Steps	*Activities*	*Program Actions*
Step 3: Develop Strategies	Identify the top barriers and benefits that emerged from the survey. If different audience segments have different barriers, decide which segment to focus on.	You decide to focus the initial pilot on those who live in single-family homes with children or young adults living in the home as they represent a large segment of the community and are more likely to have used batteries from cell phones or laptops. You create a strategy that includes information about the location of collection sites and items accepted, along with strategically placed reminders. You decide to test whether delivering the outreach through the mail is as effective as delivering it in-person.
Step 4: Pilot Test Strategies	Identify treatment and control areas. Identify metrics. Deliver the pilot strategies	You identify three geographically separated neighborhoods that are approximately 10 minutes from a collection point. You randomly assign one neighborhood to get the mailed outreach, one to get an in-person visit with outreach, and one to serve as the control. You measure weekly collection of laptop and cell phone batteries at the nearest collection sites at several points before and after the delivery of campaign materials.
Step 5: Broadly Implement	Analyze results. Identify the most cost-effective strategy. Expand the program. Develop turnkey solutions	You find that in-person outreach resulted in a 20% increase in rechargeable battery collection compared to a 15% increase from mailed outreach and a 5% increase in the control group. The cost of delivering the in-person outreach is nearly twice the cost of delivering the outreach by mail. You decide that it is most cost-effective to implement the mailed outreach campaign on a larger scale. Even though it was less effective, you can distribute it to many more households.

org for hundreds of notable examples from across the globe). There are at least four reasons that CBSM continues to gain the appeal of practitioners working to develop behavior change programs in their communities (Tabanico and Schultz, 2018).

First, CBSM is a data-driven framework, and the decisions made at each step of the program, from design to implementation, are based on the best available evidence. This is a substantial improvement over developing programs based on a hunch or historical precedence, and it offers a solid foundation for developing an effective program.

Second, the strategies developed using the CBSM framework are pilot-tested on a small scale before large-scale implementation to ensure adequate ROI. This can be a cost-saving mechanism that allows the development team to try out different approaches until they are confident that their approach is cost-effective at achieving the desired behavior change.

Third, the CBSM approach includes ongoing evaluation of scaled-up strategies to ensure that at the program's conclusion, there is data to substantiate the effectiveness (or lack thereof). This data is a valuable resource for informing the direction of subsequent outreach campaigns, justifying program funding, supporting policy change or adoption, or demonstrating compliance with mandates (e.g., mattress diversion).

Finally, the CBSM approach focuses on behavioral outcomes. Many public programs tend to focus on proxies for behavior such as engagement, likeability, attitudes, or stated intention, rather than behavioral outcomes. This tendency is due, in part, to efficiency, given that measuring behavior requires time and may not be observed immediately. Conversely, social media engagement, website visits, or self-reported behavioral intentions and attitudes can be assessed immediately. Because behavior change typically occurs later and in a different context from the intervention, it adds complexity to the evaluation process. While attitudes and intentions are easy proxies, there is ample evidence that attitudes or intentions can change without a corresponding behavior change. Timing of attitude or intention assessment also matters, such as improved correlation when intentions are measured close to the time of the behavior. In other words, behavior change resulting from an intervention can be associated with changes in attitudes or intention, but in most cases focusing only on these proxies as the primary outcome is not a sufficient substitute for measurements of behavior.

Community-based social marketing has been successfully used for nearly three decades to change a wide range of environmentally responsible behaviors, including energy conservation, water conservation, recycling, biodiversity, public transportation, waste reduction, and water quality (McKenzie-Mohr et al. 2011; see also www.cbsm.com).

In my agency, Action Research, we have successfully used the CBSM approach to address a wide range of environmental behaviors, including those in waste reduction and management. For example:

- In 2021, we worked with the Salinas Valley Solid Waste Authority to develop outreach materials for a new residential curbside organics collection program. The outreach was developed using CBSM as a program development framework. Two strategies were developed using information gathered from audience research to address key barriers and highlight benefits. The strategies were pilot-tested and compared to a control group of households who only received information about the program. During the pilot, 21 percent of households that received informational materials along with a countertop kitchen pail delivered to their doorstep placed food scraps in their green cart compared to just 10–12 percent of households who received mailed material with a voucher to pick up the countertop kitchen pail and 8 percent of households who received mailed information only with no pail (Action Research 2022a).
- In 2016, we worked with StopWaste, a public agency responsible for increasing recycling and reducing waste in Alameda County (California) to identify outreach strategies that would improve participation in an existing curbside organics collection program. The project consisted of a mail survey to identify barriers and benefits followed by testing three outreach strategies in the City of Fremont. The three strategies were (1) a mailed report showing the rates of organics recycling in their neighborhood compared to nearby neighborhoods (normative feedback), (2) a hangtag placed on trash bins (prompt), and (3) a kitchen counter pail delivered to households along with informational materials (convenience, prompt, and needed information). Three waste collection routes were assigned to receive the outreach materials and two served as controls.

Evaluation measures included pre- and post-outreach waste audits of trash bins, route-level tonnage, and a post-pilot survey of households. All three outreach strategies resulted in significant reductions in the overall trash weight, the weight of compostable materials in the trash, and the amount of food scraps in the trash compared to control routes (Phelps et al., 2017).

- In a 2014 study commissioned by Keep America Beautiful with support from PepsiCo Recycling and CBRE (Keep America Beautiful, 2015), we pilot-tested four strategies to increase recycling and reduce contamination in deskside recycling bins in commercial office settings. The four strategies were based on overcoming the barriers to recycling in the office context—primarily, convenience and information. A total of thirty-five offices in Atlanta, Houston, Boston, and San Diego were randomly assigned to one of four strategies: (1) two equal-sized recycling and trash bins at the desk; (2) recycling bin only at the desk; (3) recycling bin at the desk with "little trash" bin hanging on the recycling bin; and (4) no recycling or trash at the desk. An informational flyer was provided in all conditions. The strategies were evaluated using audits of the weight and percentages of recyclables in the recycling and trash bins, frequency of materials properly and improperly placed, and a follow-up employee survey assessing knowledge, attitudes, and behaviors related to workplace recycling. The most effective configuration was providing a recycling bin paired with a small trash bin hanging on the side. Those in the "little trash" condition increased recyclables from 65 percent to 85 percent and decreased contamination in the recycling bin by 20 percent. Employees with small deskside trash bins also demonstrated increased awareness and knowledge about recycling (Keep America Beautiful 2015).
- In Texas, we worked with the City of Fort Worth Solid Waste Services Division to improve the city's recycling program by identifying ways to increase participation and reduce load contamination. Upon project initiation, the city reported that the diversion rate was 18 percent and that 30 percent of the recycling was contaminated with nonrecyclable materials. We began by conducting a waste characterization study (cart audits) of trash and recycling carts from 415 randomly selected households across the city to identify the optimal target behaviors for a CBSM campaign. Based on the audits, we selected "place mixed paper in the recycling bin" as the primary target behavior for the campaign since it was found in 87 percent of the garbage carts and comprised 8 percent of the volume

of the cart (by weight). The waste audits were paired with 261 in-person surveys completed by the same households in the audit. The surveys were used to identify barriers and benefits to recycling six materials of interest, as well as the values, attitudes, and beliefs related to recycling and garbage. The primary barriers to recycling mixed paper included concerns about privacy and concerns about the recyclability of shiny paper and envelopes with windows. Perceived benefits centered around the environment, social responsibility, and reducing the need for landfill space. The findings from the survey directly informed the development of a strategy that consisted of attaching a rack card with a black marker to the residents' household recycling cart. The message urged residents to mark out private information using the marker and then recycle the paper. The behavior change tools incorporated into the outreach material included social norms, information, and convenience. Program messages emphasized privacy and safety through the tagline, "Keep your private information in the trash—mark out and recycle." The strategy was pilot-tested using an experimental design. Two treatment routes received the program materials. The total weight of each recycling route was collected and compared to matched routes that did not receive the program materials (control group). Unfortunately, Fort Worth could not collect sufficient data to provide conclusive results for this pilot. However, pairing household audits with barrier research related to specific materials represents a replicable model for developing targeted outreach (Tabanico and Schultz, 2018).

These examples illustrate the potential of the CBSM approach for developing programs that foster measurable behavior change. The approach is applicable to a wide range of waste-related behaviors. In the sections that follow, the development of a CBSM strategy will be illustrated from the ground up through an exercise using your own topic of interest and context as an example.

Step 1: Strategically Select Behaviors to Be the Focus of the Program

Given that environmental problems are behavior problems, they require behavioral solutions. Yet, as previously mentioned, one should not presume to know barriers, benefits, and behaviors. What is the behavior that is most important to change? Verify intuition and assumptions to ensure that the behavior selection is strategic and grounded in data. A well-known psychological phenomenon known as *confirmation bias* suggests that humans systematically seek out information that confirms our preexisting beliefs and while disregarding information that does not fit with these beliefs. Therefore, we must start program development by setting aside speculation to ensure we make decisions based on objective information.

Define Outcomes

The most critical decision prior to launching the development of a behavior change program is to determine what needs to change. Every decision about the target audience and behavior(s) should be with the end goal in mind. For example, is the goal to divert a specific material from landfilling or incineration? Is it to prevent wildlife from subjection to marine debris and inevitable entanglement by fishing gear or plastics? Or perhaps the goal is to increase the quality (i.e., decrease contamination) of collected recyclables or, further upstream, to increase purchase and consumption of products with post-consumer recycled content.

As you ponder the issues you seek to address, consider the behavior(s) and who your audience might be. If it is useful at this point, you may want to write down an outcome that you hope to achieve. In doing so, consider the available evidence of its magnitude and environmental benefit relative to other issues.

Planning Prompt for Readers

- What is your primary outcome of interest?
- What evidence supports the magnitude of the problem and the importance of this outcome relative to other issues?

Choose Relevant Sectors or Targets

Once you have determined the outcome that is most important to address and established its importance, the next step is to identify the sector that is most closely tied to the problem. For example, suppose your outcome of interest is to reduce the amount of organic material going to a landfill or incinerator, it is useful to determine what sectors are associated with the highest generation of organic material (i.e., residential, commercial, industrial, or agriculture sector). It is often useful to graphically represent the desired outcome along with a separate chart showing each sector's contribution. Choose one target market to focus on for your behavior change campaign.

Planning Prompt for Readers

- Which sectors or targets are most closely tied to the problem?
- Is your sector or target selection impacted by your mandate, jurisdiction, or other restrictions?
- Based on the above considerations, which sector will be your program focus?

List Possible Behaviors

Next, we need to look at the identified sector or target and ask specifically what these individuals are doing that contributes to the problem or outcome of interest. When listing these behaviors, make sure that the behavior meets the following criteria:

1. Non-divisible. Make sure that the behavior cannot be divided into other actions. For example, "reduce waste" or "properly recycle" can be broken down into dozens of actions based on material or context.
2. End-state. This is the behavior that has an immediate impact on the outcome. For example, purchasing a backyard composter does not reduce waste. Placing appropriate food remains in the bin and properly composting has the impact.
3. Avoid strategies. Do not list actions such as attending a workshop, visiting a website, and the like. While these are activities that you hope will foster adoption of the target behavior(s), they do not directly impact the outcome.

Note that behaviors may be one-time/infrequent (e.g., taking a used mattress to a recycling center) or repetitive/frequent (e.g., bringing a reusable bag to the grocery store).

Table 5.2 Sample Table for Listing and Defining Behaviors

#	What Is the Behavior?	End State?	Strategy?	Type
1		□ YES □ NO	□ YES □ NO	□ Repetitive □ One Time
2		□ YES □ NO	□ YES □ NO	□ Repetitive □ One Time
3		□ YES □ NO	□ YES □ NO	□ Repetitive □ One Time
4		□ YES □ NO	□ YES □ NO	□ Repetitive □ One Time
5		□ YES □ NO	□ YES □ NO	□ Repetitive □ One Time

Planning Prompt for Readers

You may choose to practice listing your behaviors in a table such as in Table 5.2. List as many behaviors as you can think of that directly impact the desired outcome. Then go back through the list and ask: Is the behavior non-divisible? Is the behavior one that is end state? Is this a strategy? Indicate also if the behavior is one-time or repetitive.

Prioritize the List of Behaviors

In many cases, there will be more than a handful of behaviors that are viable candidates for a campaign. When this is the case, we must engage in strategic efforts to prioritize the behaviors. When faced with too many choices about what behaviors to engage in, audiences will likely get overwhelmed with information and disregard the behavioral asks altogether. Efforts to prioritize behaviors are based on a consideration of four factors: impact, penetration, probability, and applicability. These factors are defined as follows:

1. Impact represents the strength of the relationship between the behavior and the outcome. Impact information should be gathered using observations or results of concrete measurements (e.g., waste audits). When this is impossible, impact information may be gathered by averaging independent ratings provided by experts, whereby each

behavior is rated relative to one another using a scale that begins at zero (no impact).

2. Penetration is the extent to which the audience is already engaged in the desired action. Penetration information is best gathered through direct observation of behavior. When this is impossible, penetration information may be gathered by surveying members of the target audience and asking them to self-report their engagement in each behavior under consideration for the program.
3. Probability represents the willingness of the audience to engage in the action. The best source of probability information comes from reviewing best-in-class programs or studies and determining what level of participation was achieved. When this information is unavailable, it is most appropriate to survey members of the target population and ask them to self-report their willingness to engage in the various behaviors under consideration.
4. Applicability is defined as the extent to which the behavior is relevant in the total population. For example, recycling used motor oil applies only to the proportion of the population that are do-it-yourself (DIY) oil changers. Similarly, the behavior of using compostable packaging only applies to the proportion of the population serviced by collection infrastructure for these items. Information about applicability may be gathered from a variety of sources including the US Census, GIS Mapping, waste collection service area data, and representative surveys.

Planning Prompt for Readers

- Given your selected outcome, resources, and available data, how would you anticipate gathering information on impact, penetration, probability, and applicability for the behaviors you are considering promoting?

Choose the Best Behavior for the Program

Once you have gathered information regarding impact, penetration, probability, and applicability, you can select your behaviors by calculating a weight that considers these factors all together. The behaviors that rise to the top of the list (have the most weight) represent the best combination of high impact, low penetration, high probability, and broad applicability. Calculate a weight for each behavior using the following formula:

$$\text{Weight} = \text{Impact} \times (1 - \text{Penetration}) \times \text{Probability} \times \text{Applicability}$$

Ideally, you should summarize the information in a table, such as in the example of Table 5.3. (Note that the numbers in the example table are fictional and intended for demonstration.) Using the earlier coffee shop example, imagine you gathered impact figures by gathering expert ratings from 0 (no impact) to 10 (maximum impact). You conducted observations at a sample of coffee shops to estimate penetration (i.e., the proportion of patrons observed engaging in the behaviors). You gathered probability information by surveying a random sample of patrons about their willingness to engage in each behavior from 0 (not at all) to 10 (definitely). Finally, you obtained applicability figures from market research estimates of the proportion of coffee shop patrons who purchase various items (hot beverages, food items, etc.).

Behaviors in the first tier (higher weights relative to others on the list) should be promoted to the barrier and benefit research phase. Additional considerations when deciding which behaviors to move forward with might include the availability of resources or control to overcome anticipated barriers, fit with your organizational mission or mandate, equity concerns, and the likelihood of the program supporting future funding or policy.

Identify and Break Down Behavior Chains

Once you have narrowed down a list of actions, you will need to determine whether any behaviors represent a behavioral chain. Often, behaviors are composed of a series of sub-actions, each of which has unique barriers. For example, the action of using a reusable bag for produce items begins with identifying suitable reusable bag(s), determining how many bags might be needed for standard produce purchases, obtaining the bag(s), placing the bag(s) in a vehicle or other location where it will be taken to the grocery store, taking the bag(s) into the store, placing products in the bag(s), and then returning the bag(s) to the proper location so that it is available for the next

Table 5.3 Sample Table for Calculating Weights to Prioritize Behaviors

Single-Use Plastic Reduction Behaviors at Coffee Shop	**Impact (0–10)**	**Penetration (%)**	**Probability (0–10)**	**Applicability (%)**	**Weight**
Refuse a plastic straw or accept a straw	8	0.20	9	0.50	28.80
Consume hot beverages without a plastic lid (don't ask for a lid)	4	0.05	7	0.70	18.62
Bring a durable, reusable bag for takeout items	7	0.05	6	0.50	19.95
Bring a durable, reusable container for takeout items	6	0.05	4	0.20	4.56
Buy food without plastic packaging	4	0.20	2	0.20	1.28

Note: Sample calculation for not using a plastic straw: 8 × (1–0.20) × 9 × 0.50 = 28.80.

use. At each step along this behavioral chain, unique barriers may impede action. To facilitate behavior change, the barriers at each step along the chain need to be addressed. It may be beneficial to unpack target behaviors into their components to ensure that the full range of barriers and benefits are adequately understood. Additionally, breaking down behavioral chains may highlight areas where the audience is currently falling out of the process, allowing you to focus programs on these points (e.g., focusing on helping people remember to take bags to the store rather than on the step of obtaining them). It is also possible that mapping the behavioral chain will help you simplify the behavior by identifying ways to take steps out of the process (e.g., providing reusable produce bags at grocery stores).

Planning Prompt for Readers

- Can any of the behaviors you are considering for your program be represented by a behavioral chain? If so, you may find it useful to draw this chain.
- What do you know about where most people are in the chain of actions? Are there points where people "fall out"?
- Can anything be done from a programmatic standpoint to simplify the number of actions that need to be taken?

Step 2: Identify Audience Barriers and Benefits

Essential to the development of an effective behavior change strategy is an objective understanding of what barriers prevent the adoption of the behavior(s) and what would motivate people to act. Once you have identified which behavior or small set of related behaviors you wish to focus on with your program, you can begin to design a research strategy to answer the following questions: (1) What does the audience perceive to be difficult or challenging with respect to engaging in the target behavior? (2) What does the audience see as benefits that might be realized if they were to engage in the behavior? (3) What are the barriers and benefits that differentiate those who presently engage in the behavior (active) from those who do not currently engage in the behavior (inactive)? (4) How do barriers and benefits differ across defined groups (e.g., renters vs. homeowners)?

Gaining an adequate understanding of barriers and benefits involves using a combination of research techniques including literature reviews, behavioral observations, focus groups, and surveys. Each method comes with inherent strengths and weaknesses. Therefore, using a combination of methods is recommended. An overview of the most common methods and a discussion of their utility follows.

Literature Reviews

The first step in conducting barrier and benefit research is to conduct a literature review. This method should be used in all instances, regardless of resource constraints, behaviors of interest, or audiences. Literature reviews involve reviewing information gathered by others to determine the range of barriers and benefits that have been uncovered concerning your target behaviors. Consider various sources in your review, including technical reports, peer-reviewed journal articles, case studies, conference proceedings, and unpublished internal reports (gray literature). The primary strength of literature reviews is that they are a low-cost and efficient method that can provide an initial objective understanding of the barriers and benefits. They can provide a solid foundation for building other research tools. On the downside, literature reviews have limited generalizability as they rarely focus on the exact same context, audiences, behaviors, or geography as your own program focus.

Observations

Unobtrusive observations of behavior as it is naturally occurring provide a great deal of information about the barriers that people face to engaging in the target behavior. It can also provide clues as to what differentiates those who engage in the behavior you are trying to promote from those who do not. For example, suppose you were interested in developing programs to reduce the number of recyclable items placed in trash containers on a university campus. You might observe that most of the improper recycling behavior is attributed to campus visitors rather than to the student and staff/faculty population. You may also observe that there are certain items that both groups commonly place in the wrong bin, suggesting knowledge barriers around the recyclability of certain items. Observations provide important information about the context, audience characteristics, and infrastructure barriers, and they are not subject to the social desirability bias inherent in other methods.

On the downside, observations can take considerable time and effort, particularly for behaviors that do not occur frequently (e.g., observing littering behavior). This method is also limited to behaviors that can be observed publicly. For example, waste audits could be conducted to identify barriers to recycling behaviors (e.g., which items are placed in incorrect bins), but they would not be useful for understanding barriers to food waste reduction behaviors during meal preparation. Observations alone will rarely provide adequate information on which to develop a strategy. However, in combination with a literature review, they can provide a solid foundation for developing questions for focus groups and survey research.

Focus Groups

Focus groups involve bringing a small group together (typically five to eight individuals) to discuss individual experiences with or perceptions of the target behavior. The discussions are typically 60 to 90 minutes and may be conducted virtually (e.g., via Zoom) or in person, as appropriate for the audience. Focus groups provide an opportunity for in-depth exploration of barriers and benefits and are particularly useful for learning about new behaviors, complex behaviors, or behaviors involving numerous steps. The purpose is to gather a range of opinions objectively and not to achieve group consensus. Separate groups should be conducted for active (those engaged in the behavior) and inactive groups (those not engaged in the behavior) so that participants feel comfortable sharing diverse opinions and experiences. For example, if someone is not yet participating in their local paint collection and recycling program, they may feel uncomfortable sharing their experiences in the presence of people participating regularly. Unlike other research methods, focus groups also provide an opportunity to capture nonverbal information, such as facial expressions and posture, that may indicate approval, understanding, or interest with respect to the behaviors being discussed. Because focus groups consist only of a few people who are typically compensated for their participation, they are not considered a representative sample. Therefore, focus group outcomes cannot be used to estimate the broader population. They are, however, instrumental in providing an in-depth understanding of the consumer experience and informing the development of relevant survey items. Focus groups can be conducted by staff within your organization with guidance from outside resources on best practices (such as Smithson, 2007), or you may choose to hire an experienced facilitator.

Quantitative Surveys

Surveys provide an excellent method for obtaining information about the target audience, including those who are not currently engaged in the behavior. Unlike focus groups, a well-conducted survey will provide information about the extent of the magnitude and extent of barriers experienced by various population segments. The barrier and benefits survey aims to obtain relevant data from a broad and representative sample of the target population. Surveys can be conducted by staff or volunteers from within your organization or partner organizations. Alternatively, you may choose the outside services of a consulting firm or local university.

As of this writing, the most prominent methods currently used to collect survey data are postal mail, web, in-person interviews, and telephone surveys (Dillman, Smyth, and Christian, 2014). When properly implemented, each method can provide high-quality data to inform the development of strategies. Determining which method to use should be based on the topic, target population, and available resources. While an in-depth discussion of the various survey methods available, along with their respective strengths and weaknesses, is beyond the scope of this chapter, there are several useful resources that may support your efforts at this step. For example, the American Association for Public Opinion Research (AAPOR), the leading professional organization for public opinion and survey research, regularly publishes white papers, articles, and other guidance on survey best practices including survey design, sampling, administration, and analysis.

Barrier and Benefit Research on a Shoestring

Budget cuts, grant deadlines, staffing shortages, or other organizational constraints often make the idea of comprehensive barrier and benefit research seem daunting. Earlier in this chapter, we established that skipping this step could result in delivering strategies that do not adequately address barriers. So what should you do when you do not have the time, staff, budget, or other resources to conduct comprehensive audience research? When this is the case, Doug McKenzie-Mohr, founder of CBSM, suggests that there are three quick and inexpensive activities you can complete that will be far superior to reverting to an information-based campaign or building a program based on assumptions (McKenzie-Mohr, 2011). First, conduct a literature review. This can be completed in as little as a few hours and will provide some outside perspective about the underlying influences surrounding your target behavior(s). Second, get out of your office, board room, Zoom room, or home office and gather information from your target audience via observations (if appropriate) and brief intercept interviews. Intercept interviews involve visiting a variety of locations frequented by members of your target population and asking for participation in a brief (3–5 minutes) survey. Be sure to visit a variety of locations to improve representativeness, interview people at random to reduce bias, and ensure participation by those who are engaged in the behavior already as well as those who are not. For example, if you are interested in developing a program to ensure proper disposal of unused paint, you may situate yourself outside of local home improvement stores in your community. If you want to develop a program to ensure proper recycling of used motor oil and filters, you may situate yourself outside an auto parts store. Do not worry if you don't have the time or resources to construct a survey. It is sufficient in these cases to ask just the following three critical questions, where X represents your target behavior:

1. Do you currently do X?
2. What do you perceive to be difficult or challenging about doing X?
3. What do you perceive to be beneficial or rewarding about doing X?

Planning Prompt for Readers

Consider choosing one behavior that you wish to target with a behavior change program and answer the following questions:

- What resources might you consider in a literature review?
- Is the behavior observable? If so, where might you go to observe people engaged in the behavior?
- Considering your organizational resources, time, behavior, and other contextual constraints, what other research might you conduct to identify barriers and benefits?
- What additional support, information, or resources might you need to carry out this research?

Step 3: Develop a Behavior Change Strategy

Decades of research and theory development in psychology have provided tools for changing behavior. While behavioral scientists have been aware of these findings for nearly 100 years, policymakers and public relations firms have only recently begun incorporating them into their programs targeting behavior change (Tabanico, Schmitt, and Schultz, 2015). Although these tools are readily accessible, understanding the science behind each tool is essential. There are specific conditions under which each tool will work most effectively, and some tools can even be counterproductive if used without appropriate knowledge of their theoretical underpinnings (Osbaldiston and Schott, 2012). For example, when used incorrectly, social norms messages could inadvertently produce behaviors counter to what is intended, such as when descriptive and injunctive norms are at odds in a message (Cialdini, 2003). Similarly, combining incentives (e.g., prizes or rebates) with commitment strategies (e.g., pledges) can undermine the tool by inhibiting the formation of an intrinsically motivated self-concept. While a comprehensive review of the principles and techniques that might be used to motivate behavior change is beyond the scope of this chapter, I've included a summary of eight common tools below. Each summary includes a description of its function, suggestions for when and how the tools should be utilized, and concrete examples of their application.

Make It Easy with Convenience

It has been well established that making behavior more convenient (such as by reducing the number of steps, making the desired behavior the default, or reducing the uncertainty of participation) can increase participation. For example, in work with the County of San Diego, the California Department of Transportation, and others across the country, we have consistently found an association between the presence of adequate infrastructure for the proper disposal of trash or cigarette butts and decreased littering rates (see, e.g., Schultz et al., 2011). Importantly, when invoking convenience, it is not enough to simply advertise and try to convince people of the ease of participation.

You must implement strategies that make the behavior easier to do. Messages alone are unlikely to be effective unless the underlying structural barriers are also addressed.

When addressing convenience in your program strategies, another consideration is to seek ways to make the undesired behavior less convenient. For example, in our work with Keep America Beautiful (2015) to promote office recycling and reduce the number of recyclables going into the trash, the most effective strategy made recycling beverage containers and other items convenient by placing a recycling bin at the deskside. At the same time, hanging a small trash can attached to the side of the recycling bin served to make the incorrect behavior (placing recyclable material, such as beverage containers, into the trash) more difficult due to the small size of the container.

Planning Prompt for Readers

As a further practice, you may find it useful to think of one of the behaviors that you are considering promoting that may be somewhat inconvenient for the audience to engage in.

- How might you make this behavior more convenient for your audience, such as by decreasing barriers or simplifying steps?
- Is it also possible to make the undesirable alternative less convenient by increasing the barriers?

Enhance Self-Perceptions through Commitment

Humans have an innate desire to remain consistent in their thoughts and actions. To maintain this consistency, we often change our beliefs or attitudes to match our behaviors. Similarly, we will alter subsequent behaviors to be consistent with our earlier actions. This principle is most invoked in program strategies by obtaining an initial commitment (either verbal or written) to engage in the target behavior or asking for engagement in a small first step (put a small sticker in the window) and following up with a more significant request. Commitment strategies most commonly take the form of signing a pledge card, but it might also be invoked through stickers (e.g., an "I put hazwaste in its place at the regional HHW depot" decal placed on the garbage bin), or through online posts (e.g., changing a profile picture to show commitment to proper disposal of HHW).

Although commitments can take a variety of forms, research suggests that they are most effective when they are written (e.g., signing a pledge), made public (e.g., posted at the store entrance), and durable (e.g., the posting is in place for a long time). Importantly, commitments must be voluntary and not coerced through social pressure or incentives. Therefore, commitments tend to be most appropriate in situations where an individual already has some preexisting motivation to engage in the action but has not yet engaged (Lokhorst et al., 2013).

Commitments are a relatively low-cost program element and are relatively easy to implement. They also come with two additional benefits. First, when used effectively, commitments provide an opportunity for people to build self-perceptions around the desired set of behaviors (e.g., I am the kind of person who donates usable goods) which may, over time, build identities that support engaging in related actions (Lauren et al., 2019). Additionally, public displays of commitments (particularly when grouped together), can foster social diffusion and social norms (providing three tools for the price of one!). Social norms and social diffusion are discussed in detail in the next sections.

In a recent program developed for the County of Orange in California, youth skateboarders were asked to sign a commitment to throw their trash away before leaving the park. These signed commitments were grouped together and placed on a large poster at the entrance to the park. The efficacy of the commitment strategy was evaluated through trash counts at several points before and after the program. The rate of trash accumulation in the areas immediately adjacent to the park decreased substantially after the program's implementation (Action Research, 2022b).

Planning Prompt for Readers

Consider another behavior you might promote with an audience that is already motivated to take action.

- How might you obtain a written commitment?

- How might you make the commitment public and ensure it is long-lasting (durable)?
- Are there ways to group the commitments together in a common area?

Leverage the Power of Social Norms

Humans are social animals, and as a result, we use the behavior of others as a guide for our own actions. Seeing other people doing something (such as utilizing a program), or even having the perception that other people are doing it, legitimizes the behavior and increases its frequency. Similarly, deviating from a norm can prompt feelings of pressure to conform. The most common way to integrate social norms into a behavior change strategy is to highlight the many people who already do the behavior (descriptive norm) and approve of the behavior (injunctive norm). It may also be invoked by making a behavior public (e.g., household battery collection bins at the store). It is important to recognize that individuals tend to conform to the perceived norm for the group, even if the group is engaged in the behavior opposite to the intended behavior. Therefore, program planners must avoid conveying a norm that most people are not engaged in or do not approve of the behavior. If the behavior is not normative, other tools (such as social diffusion, discussed next) would be a better choice.

In a recent example, Action Research worked with the Salinas Valley Solid Waste Authority to promote a residential food scrap collection program. Because the program was new, the descriptive norm for the behavior was low (most people were not participating). Focusing the program on the descriptive norm would have been an error. Messages conveying that "no one is participating, but you should" would likely backfire by suggesting that participation was not the norm. Therefore, in this case we focused the program materials on the injunctive norm by highlighting survey findings that most residents supported the behavior. The flyer stated: "79% of residents think it is important to reduce waste going to the landfill by placing food scraps in the green cart."

Another effective way to use social norms is through normative feedback whereby audiences are provided information about their own behavior compared to that of similar others. For example, in our own work with the County of Alameda in California (described earlier in this chapter), we used normative feedback as one of three pilot strategies for increasing participation in a curbside organics collection program. We used a similar mailed feedback strategy in working with the State of Massachusetts to reduce summer water use in several communities drawing from the Ipswich River. The "Healthy Lawn, Happy Summer" campaign provided residents with targeted information and feedback about their water use relative to similar homes in the community. Pilot tests conducted in the summers of 2018 and 2019 demonstrated statistically significant reductions in water use (up to 128 gallons/day per household) as a result of the feedback intervention. See https://www.mass.gov/forms/healthy-lawn-happy-summer-toolkit.

Social norms messages can be used across a wide range of behaviors (Nolan, 2021). Because they operate by invoking social pressures, normative messages can work well for people who are not already motivated to engage in the behavior, as people will tend to shift their behaviors to align with the group norm. In the energy domain, for example, normative feedback interventions have successfully produced behavior changes even for those who do not pay for the energy costs (Bator et al., 2019). Unlike commitments which tend to work best for those who are already motivated but have not followed through, social norms strategies are best used for target audiences with low motivation.

Planning Prompt for Readers

Consider a behavior related to your outcome of interest that is relatively common and approved of by members of the target audience.

- Is the behavior relatively common? If so, how might you use a descriptive norm to highlight that other people are engaged in the behavior? Is it feasible to provide people with feedback about their level of engagement in the behavior compared to similar others?
- If the behavior is not common, do most people approve of engaging in the behavior? If so, how might you use an injunctive norm to highlight this social approval?

Start Something New with Social Diffusion

Behavioral research has consistently demonstrated that individuals are more likely to adopt a new sustainable behavior (e.g., using reusable products) when they have friends or family members who have already adopted the action. This process by which new innovations spread through social networks is termed *social diffusion* (also called *diffusion of innovations*). Despite a great deal of evidence that social diffusion can spur a wide range of actions, from installing programmable thermostats to drought-tolerant landscaping, it has been generally absent from many efforts aimed at promoting resource sustainability. Social diffusion is most likely to occur when the desired behavior is visible in the community (e.g., using reusable shopping bags or eating with reusable utensils at the school cafeteria). The visibility of the behavior demonstrates acceptance of the behavior by peers and encourages conversations about the behavior. Effective social diffusion strategies begin by identifying well-known and well-respected members of the target audience and obtaining a commitment from them to engage in the behavior. For example, in one of my very early CBSM consulting projects (in 2006), social diffusion strategies proved to be an effective approach to encourage Tennessee farmers to use an industrial byproduct from a fossil fuel power plant as a soil amendment on their farms. We began by identifying farmers in two regions of the state who were identified by their peers and local agricultural agents as being well-known and well-respected in the local farming community. These farms agreed to commit to using the industrial byproduct (FGD gypsum) on a small plot of their land. The results were then presented and discussed at a meeting with other farmers in the region.

In a more recent application, Action Research used a block leader model to address irrigation runoff in a single neighborhood within the City of Encinitas, a coastal California community. The block leader model used social diffusion in two ways. First, a community barrier and benefits survey included additional items to identify a well-known and well-respected individual in the neighborhood who was already engaging in the target behavior (converting landscape sprinklers to drip irrigation). The neighborhood survey overwhelmingly pointed to a single individual who subsequently agreed to host a demonstration at her home. Second, the approach facilitated conversations during a two-hour event during which homeowners could observe and participate in a demonstration identifying appropriate areas for drip irrigation and show how to install the equipment.

Planning Prompt for Readers

Consider the many behaviors that could be targeted for your outcome of interest.

- Are any of the behaviors on your list new or uncommon?
- How might you identify and ask well-known and well-respected individuals in the community to commit to the behavior? How would you further encourage conversations between these individuals and others in the community?

Address Forgetfulness with Prompts

In many cases, the barrier and benefit research reveals that people are not engaging in the desired behavior (e.g., placing recyclable items in the proper bin) because they simply forget. When this is the case, a useful tool is to provide a prompt (visual or auditory aid) close in space and time to where the desired behavior is to occur. For example, stickers on recycling bins that visually indicate acceptable recyclable materials have significantly reduced contamination. In our own work with Salinas Valley Recycles, residents were given a food scraps collection container to place on their kitchen counter. The container included hot stamp impressions of what items could (and could not) be placed into the green cart (see Figure 5.2). The kitchen pail also addressed convenience by providing a simple way to collect scraps and carry them to the curbside organics collection cart.

Sometimes, forgetting to engage in a repetitive action can be addressed through a technical solution (such as motion sensors for lighting). However, for many behaviors in the waste domain, carefully placed prompts will often be necessary to remind people to engage in the action. To be effective, prompts should be simple (self-explanatory), positive (stating what to do rather than

Figure 5.2 Kitchen pail prompting proper placement of various food items.

Source: Courtesy of Salinas Valley Solid Waste Authority.

what not to do), noticeable, and placed in close proximity to the behavior. Beyond that, prompts may take many forms including text messages, geo-fenced phone notifications, stickers, hang tags, and window clings.

Planning Prompt for Readers

Consider the many behaviors that could potentially be targeted for your outcome of interest.

- Are any of the behaviors repetitive actions that the audience might forget to engage in?
- How might you make use of a prompt placed in close proximity to the behavior?

Create New Habits by Building on Existing Ones

Moving the needle on our most pressing environmental problems will ultimately require many people to engage in pro-environmental behaviors repeatedly over an extended period. Therefore, when designing behavior change strategies, it is helpful to consider ways to encourage the development of good environmental habits (and break bad ones). Many of the activities we engage in daily happen outside of our conscious awareness—they are habitual. We rely on habits to navigate our daily lives, saving us from taking time and brainpower to think through every action. For many people, bad environmental habits came about over time as a matter of convenience. For example, it takes more effort to bring a reusable water bottle or grocery bag when leaving the house than to rely on the convenience of the supplied plastic option. Habits happen automatically and outside of conscious awareness, so they are very difficult to break—even when knowledge and motivation support a change. While breaking old habits is hard, the good news is that they can be replaced with good habits that also die hard. Habits are essential context-response associations that develop as behaviors are repeated in daily life. Creating new habits requires three components: (1) repetition, (2) context cues, and (3) rewards. Interventions that focus on these elements are likely to be successful by building strong habits (Wood and Neal, 2016).

One way to build opportunities for habit formation into your strategies is to use the concept of *habit stacking*. Habit stacking involves leveraging existing cues by ordering a new behavior to follow an established habit. In this way, the existing habit serves as a prompt or cue for the new behavior, which is likely to be repeated as part of the new habit string. For example, the County of San Diego in California implemented a neighborhood-based program to reduce instances of irrigation water flowing into the storm drain system. The primary target behavior was encouraging residents to check their irrigation systems for broken or misaligned sprinklers and then make repairs. During the barrier and benefit research, residents reported that they were unaware if their sprinklers needed repair. Irrigation cycles typically ran in the late evening or early morning hours, and residents did not regularly check their systems. Residents reported regularly mowing their lawns (about twice a month or more often). These conditions presented an opportunity to create a strategy in which a habitual behavior (regularly mowing the lawn) was paired with a new behavior (checking for and repairing broken sprinklers). The resulting "Check After You Mow" program has been deployed in several San Diego communities and has consistently demonstrated reductions in irrigation runoff.

Planning Prompt for Readers

Consider whether any of the behaviors you are considering for a program occur on a repetitive basis.

- Are there opportunities to increase the frequency with which they practice the behavior?
- Is there another behavior (existing habit) that might serve as a cue for a new behavior?

Enhance Follow Through with Goal Setting

Most of us will set numerous personal or professional goals in our lifetime (e.g., the infamous New Year's resolutions), but these stated goals are rarely achieved. Fortunately, a growing body of research indicates that we are more likely to achieve our goals if we create a specific (when and where) action plan for achieving that goal, an *implementation intention* (Gollwitzer, 1999; Orbell and Sheeran, 2000). When an audience is interested in engaging in a particular action, ask them to create an implementation intention stating details such as when and where they intend to engage in the behavior. Creating this simple implementation encourages the individual to engage in some of the upfront cognitive work that is ultimately required for the behavior change. In 2021, Action Research worked with the City of Palmdale to develop a program to enhance the uptake of a free bulky item pickup service offered to residents living in multifamily homes. The program was intended to address growing issues with the illegal dumping of items such as furniture and mattresses in areas adjacent to multifamily properties. The multifamily outreach was composed of a program landing page with information about disposal

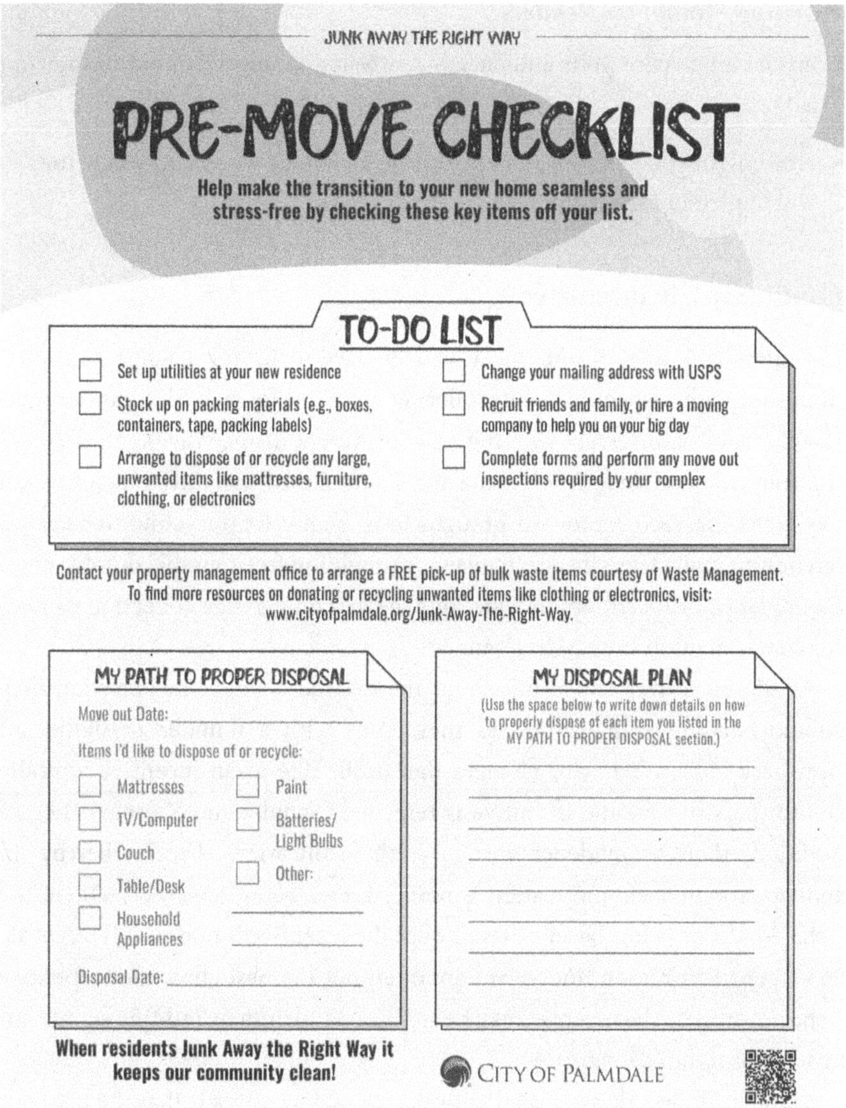

JUNK AWAY THE RIGHT WAY

PRE-MOVE CHECKLIST

Help make the transition to your new home seamless and stress-free by checking these key items off your list.

TO-DO LIST

- [ ] Set up utilities at your new residence
- [ ] Stock up on packing materials (e.g., boxes, containers, tape, packing labels)
- [ ] Arrange to dispose of or recycle any large, unwanted items like mattresses, furniture, clothing, or electronics
- [ ] Change your mailing address with USPS
- [ ] Recruit friends and family, or hire a moving company to help you on your big day
- [ ] Complete forms and perform any move out inspections required by your complex

Contact your property management office to arrange a FREE pick-up of bulk waste items courtesy of Waste Management. To find more resources on donating or recycling unwanted items like clothing or electronics, visit: www.cityofpalmdale.org/Junk-Away-The-Right-Way.

MY PATH TO PROPER DISPOSAL

Move out Date: ____________

Items I'd like to dispose of or recycle:

- [ ] Mattresses
- [ ] TV/Computer
- [ ] Couch
- [ ] Table/Desk
- [ ] Household Appliances
- [ ] Paint
- [ ] Batteries/ Light Bulbs
- [ ] Other: ____________

Disposal Date: ____________

MY DISPOSAL PLAN

(Use the space below to write down details on how to properly dispose of each item you listed in the MY PATH TO PROPER DISPOSAL section.)

When residents Junk Away the Right Way it keeps our community clean!

CITY OF PALMDALE

Figure 5.3 Using implementation to address proper disposal of bulky items.

Source: Courtesy of City of Palmdale.

services, informational door hangers, and a move-out checklist flier. The flier was designed to leverage a time that was commonly associated with disposing of bulky items (move out) and help residents create a plan (implementation intentions) for disposing of these items (see Figure 5.3).

Planning Prompt for Readers

Consider a behavior your audience is motivated to engage in but has not yet acted on.

- How might you help the audience create a plan for when and where they will implement the action?

Cautiously Use Incentives

Incentives are one of the most widely used behavior change tools. As discussed with respect to habits, offering a reward for behavior can increase its frequency. Conversely, offering a fee or fine (a disincentive) can decrease the frequency of undesirable behavior. Examples of incentives include cash payments for used motor oil brought to a facility for recycling, rebates on environmentally friendly alternatives, or providing free reusable durable bags. Examples of disincentives include tonnage fees on garbage collection or fines for contamination in recycling bins.

While incentives can bring about the desired changes in behavior, they should be used with caution as they come with a number of significant drawbacks. First, behavior changes that occur due to an incentive typically do not persist once the incentive is removed (Schultz and Kaiser, 2012). In addition, there is evidence for an "over-justification effect" whereby an individual's intrinsic motivation is reduced after being rewarded. Often, the behavior drops below baseline levels once the incentive is removed (Deci et al., 1999). For this reason, incentives are often not the best choice for repetitive behaviors unless the incentive can be sustained indefinitely (adding significant costs to program delivery).

Another drawback associated with the use of incentives is that the behavior changes brought about by using incentives are unlikely to spill over to other related behaviors (Schultz, 2014). For example, offering an incentive for residents to recycle a used mattress will not likely spill over to other responsible disposal actions, such as recycling old compact fluorescent lightbulbs. Given the drawbacks associated with using incentives, they are best used in behavior change programs only when the cost is identified as a primary barrier to the action.

Planning Prompt for Readers

Consider whether any behaviors on your list might have significant cost barriers associated with them.

- Do you have the program resources to provide incentives?
- What might be some potential consequences of removing the incentive if funds run out?

Capture Attention with Effective Communication

At this point in the CBSM planning process, you have spent a great deal of time identifying the optimal behavior, understanding your audience, and selecting behavior change tools that best persuade your audience to change. But all this effort will have been wasted if the audience does not pay attention to your program message. Attracting attention alone is not enough to change behavior, but it is a critical first step, so how can you make your audience pay attention? Below are a few ways to help your message stand out in a world of information overload.

1. **Use vivid and concrete messages.** Many environmental issues are abstract, and the impacts are not always visible in our daily lives. Instead of using technical language and numbers, use vivid imagery and language that make the impacts more tangible (Gonzales, Aronson, and Costanzo, 1988). For example, imagine you are trying to describe the amount of single-use plastic waste annually generated within a community. Rather than simply saying how many tons of waste are produced, you might describe it in terms of how it would fill the local football stadium instead or how many times it would fill an Olympic-sized pool.
2. **Make it personal.** Messages that use person-to-person communication or that are tailored to a specific audience's needs are considerably more influential than passive media messages (Tabanico and Schultz, 2018).
3. **Use credible messengers.** The most influential messengers are those who have both expertise and trustworthiness (Cialdini, 2020). It is often useful to identify credible messengers as part of the barrier and benefit research process.

4. **Use appropriate message frames.** It is tempting to focus exclusively on the positive outcomes one might gain from participating in the desired behavior. However, it turns out that messages that are framed in terms of potential loss from inaction are often more influential than those emphasizing potential gains from acting (McKenzie-Mohr, 2011).
5. **Highlight local impacts.** In many cases, our target audiences do not personally experience the effects of the global environmental problems we are trying to address. To overcome this challenge, highlight the impacts that these problems have on a local level such as impacts on local wildlife or property values. Also highlight local solutions (CRED, 2014).
6. **Emphasize identity and self-perception.** Tap into the audience's desire for consistency by highlighting the ways in which the behavior aligns with the identities and priorities they already have (e.g., being a parent, dog owner, vegan).

Putting It All Together

To this point, you have been introduced to methods and best practices for selecting behaviors, conducting audience research, selecting appropriate behavior change tools, and creating effective messages. It is now time to pull this together into the development of one or more behavior change interventions (strategies).

Create a Strategy Table

The most helpful way to begin conceptualizing a program design is to develop a strategy table that lists the research findings, associated tools, potential channels, and possible messengers. Keep in mind that every program element has a cost. Using a strategy table will help to ensure that each of your program elements ties clearly to the audience research. Additionally, a strategy table will help you to identify ways in which multiple barriers or benefits can be addressed with a single tool. Diligence in using this planning approach results in a lean program design without excess (potentially costly) program elements that do not directly serve the behavior change. Your strategy table should list

Table 5.4 Sample of a Partial Strategy Table for Multifamily Bulky Waste Item Disposal

Barriers/Benefits	Tools	Strategy Options
Lack of Knowledge. Residents are unaware of what items are accepted.	Communication	Mailer In-Person Event
Structural Barriers. Collection facility hours are limited to weekdays.	Convenience	Offer extended or weekend hours Pickup service Event
Forget. Residents forget to take unwanted items that are stored in the garage.	Prompt	Text reminders Stickers Door hangers
Low Motivation. Residents are generally not motivated and believe that most people don't visit the collection facility.	Social Norms Commitment	Testimonials Signed pledges

the key research findings (barriers and benefits) along with the tools that could best be used to address them. You might also include ideas for different ways these tools could be operationalized. Table 5.4 provides an example of what a strategy table might look like for encouraging residents to take bulky waste products to a local collection facility.

Pre-Test

After developing your strategy table, you may have several different ideas for structuring one or more strategies. Whenever possible, it is often useful at this step to conduct some pre-testing of program elements or messages to ensure that the audience can understand the messages and that they resonate as intended. Pre-testing might be done through focus groups, intercept interviews, or web surveys. These forums provide an opportunity for you to show drafts of program materials, messages, and other elements before they are fully produced. For example, you are planning to motivate college students to use reusable utensils by asking them to have their photo taken and to sign a commitment and then post their photo commitment on a kiosk at the entrance to the cafeteria, but you discover that some students are unwilling to have their photo taken. In that case, you may need to adjust your strategy before pilot testing.

Step 4: Pilot Test the Strategies

Pilot testing provides an opportunity for you to test out one or more strategies to identify which are most cost-effective at changing behavior. The pilot will also allow you to identify and address any problems before broadly implementing the campaign across the community. The size, duration, and rigor of a pilot may depend on several factors that are outside the scope of this chapter. However, the following best practices should be used as guidelines:

1. **Use treatment and control groups.** To ensure that the behavior changes you observe are the result of your intervention and not other factors (e.g., weather, news events, other campaigns), be sure to include a control group that does not get the intervention. The control group serves as a base case (what behavior change would have happened in the absence of your intervention). You may choose to test a few different variations of your program (e.g., one that includes commitments and one that includes an incentive). In this case, you would designate multiple treatment groups and compare their respective outcomes to the control group.
2. **Use random assignment to condition.** To have the most confidence in your pilot results, you will want to ensure that the group that gets your intervention is as similar as possible to the group that does not get the intervention (control). Random assignment to condition eliminates preexisting differences by ensuring that every participant has an equal likelihood of being in one of the treatment groups or the control.
3. **Revise the pilot until it is effective.** If none of the strategies produce a measurable behavior change, or if the level of behavior change is too small to justify the cost, then make adjustments to the program and re-pilot the strategy before implementing across the community. Only move to broad-scale implementation when you have evidence supporting a cost-effective strategy.

If the strategies that you tested did not work, it is important to note that this should not be interpreted as "CBSM did not work." One unfortunate misconception that I have observed in my work with dozens of agencies who have applied or tried to apply CBSM to their behavior change programs is the misconception that CBSM is a strategy. This mindset often results in the conclusion, "We tried CBSM, and it didn't work." The error here is that CBSM is a process and not a strategy.

There are at least four reasons why strategies developed using CBSM are not effective: (1) the strategies did not adequately address barriers, (2) they did not adequately enhance motivation, (3) they did not effectively reach or capture the attention of the audience, or (4) the outcome measurements were not sensitive enough to detect change. Therefore, it is likely an error to state that the process failed. During the pilot testing phase (Step 4), strategies are tested until they are effective. Thus, CBSM is inherently designed to ensure success. The process works, but not every strategy will work the first time out of the gate. Medical researchers conduct numerous clinical trials and spend significant time researching before they identify the specific combination that effectively manages symptoms or cures an ailment. The same is true for developing effective strategies to change behavior.

Step 5: Broadly Implement the Most Effective Strategies

When pilot testing has adequately demonstrated that behavior change can be achieved in a cost-effective way, it may be broadly implemented across the community.

Scaling Up

When making decisions about scaling up the program, you must retain all of the original elements of the pilot program. For example, if the pilot included commitments publicly posted in a neighborhood newsletter, you cannot change to a web-based commitment strategy for the broadscale rollout. However, you may choose to supplement the pilot with advertising in local media or social media—strategies that would not have been desirable in the pilot testing phase.

Developing Turnkey Solutions

Effective strategies can be packaged into toolkits that can be implemented by other organizations or agencies who are working in a similar context.

For example, Natural Resources Canada provides access to a free toolkit for those wishing to implement anti-idling campaigns in their communities (Natural Resources Canada, n.d.). Similarly, the Urban Sustainability Directors Network developed two separate toolkits based on the success of pilots encouraging residents to wash clothes in cold water and encouraging municipal employees to turn off computers at the end of the workday (USDN, 2013).

Ongoing Evaluation

Gather baseline measurements prior to broadscale implementation, and then plan to continue to evaluate outcomes at various time points to ensure that the strategy is continuing to have an impact. It is not uncommon for behavior to change immediately and then trail off after some period. This may indicate that there are new barriers impacting the audience, competing motivations, or that the strategy was not in place long enough to establish habits. Ongoing evaluation will provide information you need to fine-tune and adjust your program to ensure long-lasting behavior change.

Getting the Word Out

Although it is tempting to stop your work at the point of implementing a successful strategy, I would encourage you to take the time to write up the results so that others can learn from your experience. Consider sharing your results (successes and failures) at professional meetings, with other producer responsibility organizations, social marketing or behavioral science conferences, web-based forums, or in peer-reviewed journals.

Conclusion

Every waste reduction or disposal program relies on human behavior. Unfortunately, most programs are based on faulty assumptions about human behavior. They assume that lack of behavior results from lack of knowledge: "If people only knew about the program, surely they would use it." If this were true, effective outreach materials would simply need to educate consumers about the program. As a result of these assumptions, we continue to see thousands of print, radio, social media, and digital advertisements nationwide that are intended to inform residents about the program or sometimes raise awareness about the seriousness of the issue.

Unfortunately, the assumption that lack of action results from lack of knowledge often turns out to be false. While lack of knowledge can be a barrier to action, it is not sufficient by itself to motivate behavior. Because of this, many existing outreach efforts fail to motivate behavior change.

Over the last few decades, a large body of evidence has emerged showing that behavioral science can help to solve a wide range of pressing environmental problems. Consequently, government and nongovernmental agencies are increasingly applying behavioral insights as a supplement to traditional approaches for shaping public behavior. In just the last decade, the application of behavioral science in the public sector has increased astronomically, and a number of governments around the world have formed "nudge units" or "behavioral insights teams" to systematically incorporate behavioral science into a wide range of public programs.

Community-based social marketing is a data-driven process that effectively brings together behavioral science, social marketing, and audience research in a way that is useful for practitioners, providing an effective and viable alternative to information-based campaigns that have proliferated in the public sector. The CBSM approach uses data from the target population to inform program and outreach material development. CBSM works by focusing on a specific behavior, surveying the target population to establish a base rate and barriers to the behavior, developing an intervention or making structural changes that directly address the barriers, and collecting data to evaluate the efforts to promote behavior change. The approach has been successfully used in many areas including waste reduction, and it offers a promising alternative for practitioners looking for an effective way to improve the success of their programs.

References

Action Research. 2022a. *Community-Based Social Marketing to Create a Food Scraps Collection Program*. Final report submitted to Salinas Valley Solid Waste Authority (May 2022).

Action Research. 2022b. "Tricks to Trash Cans." Presentation at the Annual Conference of the California Stormwater Quality Association (CASQA). October 2022.

Bator, R. J., K. Phelps, J. Tabanico, P. W. Schultz, and M. L. Walton. 2019. "When It Is Not about the Money: Social Comparison and Energy Conservation among Residents Who Do Not Pay for Electricity." *Energy Research & Social Science* 56: 1–8.

Burgess, J., C. M. Harrison, and P. Fillius. 1998. "Environmental Communication and the Cultural Politics of Environmental Citizenship." *Environment and Planning A: Economy and Space* 30 (8): 1445–60.

Casagrande, S. S., Y. Wang, C. Anderson, and T. L. Gary. 2007. "Have Americans Increased Their Fruit and Vegetable Intake? The Trends between 1988 and 2002." *American Journal of Preventative Medicine* 32 (4): 257–63.

Center for Research on Environmental Decisions, ecoAmerica. 2014. *Connecting on Climate: A Guide to Effective Climate Change Communication*. New York: Columbia University; Washington, DC: ecoAmerica.

Christiano A., and A. Neimand. 2017. "Stop Raising Awareness Already." *Stanford Social Innovation Review* 15 (2): 34–41.

Cialdini, R. B. 2003. "Crafting Normative Messages to Protect the Environment." *Current Directions in Psychological Science* 12: 105–9.

Cialdini, R. B. 2020. *Influence, New and Expanded: The Psychology of Persuasion*. 225–9. New York: HarperCollins.

Deci, E. L., R. Koestner, and R. M. Ryan. 1999. "A Meta-analytic Review of Experiments Examining the Effects of Extrinsic Rewards on Intrinsic Motivation." *Psychological Bulletin* 126: 627–68.

Dillman, D. A., J. D. Smyth, and L. M. Christian. 2014. *Internet, Phone, Mail, and Mixed Mode Surveys: The Tailored Design Method*. 4th ed. Hoboken, NJ: John Wiley and Sons.

Gollwitzer, P. M. 1999. "Implementation Intentions: Strong Effects of Simple Plans." *American Psychologist* 54: 493–503.

Gonzales, M. H., E. Aronson, and M. A. Costanzo. 1988. "Using Social Cognition and Persuasion to Promote Energy Conservation: A Quasi-experiment." *Journal of Applied Social Psychology* 18: 1049–66.

Hornik, R., L. Jacobsohn, R. Orwin, A. Piesse, and G. Kalton. 2008. "Effects of the National Youth Anti-Drug Media Campaign on Youths." *American Journal of Public Health* 98 (12): 2229–36.

Keep America Beautiful. 2015. "Keep America Beautiful Announces Results of 'Recycling at Work' Research Study." Press release (April 28, 2015). https://www.prnewswire.com/news-releases/keep-america-beautiful-announces-results-of-recycling-at-work-research-study-300073351.html (accessed June 11, 2025).

Lauren, N., L. D. G. Smith, W. R. Louis, and A. J. Dean. 2019. "Promoting Spillover: How Past Behaviors Increase Environmental Intentions by Cueing Self-Perceptions." *Environment and Behavior* 51 (3): 235–58.

Lokhorst, A. M., C. Werner, H. Staats, E. van Dijk, and J. L. Gale. 2013. "Commitment and Behavior Change: A Meta-analysis and Critical Review of Commitment-Making STrategies in Environmental Research." *Environment and Behavior* 45 (1): 3–34.

McKenzie-Mohr, D. 2000. "Promoting Sustainable Behavior: An Introduction to Community-Based Social Marketing. *Journal of Social Issues* 56 (3): 543–54. https://doi.org/10.1111/0022-4537.00183.

McKenzie-Mohr, D. 2011. *Fostering Sustainable Behavior: An Introduction to Community-Based Social Marketing.* 3rd ed. Gabriola Island, BC: New Society.

McKenzie-Mohr, D., and P. W.Schultz. 2014. "Choosing Effective Behavior Change Tools." *Social Marketing Quarterly* 20 (1): 35–46.

McKenzie-Mohr, D., N. Lee, P. W. Schultz, and P. Kotler. 2011. *Social Marketing to Protect the Environment: What Works*. Thousand Oaks, CA: Sage.

McKenzie-Mohr, D., L. S. Nemiroff, L. Beers, and S. Desmarais. 1995. "Determinants of Responsible Environmental Behavior." *Journal of Social Issues* 51: 139–56.

Natural Resources Canada. n.d. "A Ready-Made Vehicle Idling Campaign." *Natural Resources Canada*. https://oee.nrcan.gc.ca/transportation/idling/material/campaign-resources.cfm?attr=8 (accessed June 11, 2025).

Nolan, J. M. 2021. "Social Norm Interventions as a Tool for Pro-climate Change." *Current Opinion in Psychology* 42: 120–5.

Opinion Dynamics Corporation. 2010. *PY2006–2008 Indirect Impact Evaluation of the Statewide Marketing and Outreach Programs*. Report for the California Public Utilities Commission. Oakland, CA: Opinion Dynamics Corporation.

Orbell, S., and P. Sheeran. 2000. "Motivational and Volitional Processes in Action Initiation: A Field Study of the Role of Implementation Intentions." *Journal of Applied Social Psychology* 30: 780–97.

Osbaldiston, R., and J. P. Schott. 2012. "Environmental Sustainability and Behavioral Science: Meta-analysis of Proenvironmental Behavior Experiments." *Environment and Behavior* 44 (2): 257–99.

Phelps, K., L. Large, W. Schultz, and J. Ettlinger. 2017. "Keeping Compostables Out of the Trash." *Biocycle*. November 2017, 16–18.

Schultz, P. W. 2002. "Knowledge, Education, and Household Recycling: Examining the Knowledge-Deficit Model of Behavior Change." In *New Tools for Environmental Protection: Education, Information, and Voluntary Measures*, edited by T. Dietz and P. C. Stern, 67–82. Washington, DC: National Academy Press.

Schultz, P. W. 2011. "Conservation Means Behavior." *Conservation Biology* 25 (6): 1080–3.

Schultz, P. W. 2014. "Strategies for Promoting Proenvironmental Behavior: Lots of Tools but Few Instructions." *European Psychologist* 19 (2): 107–17.

Schultz, P. W., and J. Tabanico. 2008. "Community-Based Social Marketing and Behavior Change." In *Handbook on Household Hazardous Waste*, edited by A. Cabaniss 133–57. Lanham, MD: Government Institutes.

Schultz, P. W., R. Bator, L. B. Large, C. M. Bruni, and J. J. Tabanico. 2011. "Littering in Context: Personal and Environmental Predictors of Littering Behavior." *Environment and Behavior* 45 (1): 35–59.

Simis, M. J., H. Madden, M. Cacciatore, S. K. and Yeao. 2016. "The Lure of Rationality: Why Does the Deficit Model Persist in Science Communication?" *Public Understanding of Science* 25 (4): 400–14.

Smithson, J. 2007. "Using Focus Groups in Social Research." In *The Sage Handbook of Social Research Methods*, edited by P. Alasuurtari, L. Bickman, and J. Brannen, 356–71. Thousand Oaks, CA: Sage.

Tabanico, J., and P. W. Schultz. 2018. "Designing Outreach Programs That Change Behavior." In *Handbook on Household Hazardous Waste*,edited by A. Cabaniss, 2nd ed, 137–66. Lanham, MD: Bernan Press.

Tabanico, J., J. Schmitt, and P. W. Schultz. 2015. "Driving Change: The Role of Theory in Social Marketing." In *Handbook of Persuasion and Social Marketing*, edited by D. Steward, 119–54. Santa Barbara, CA: Praeger.

Urban Sustainability Directors Network. 2013. "Energy: Innovation Lab Concept Developing and Prototyping." https://www.usdn.org/products-energy.html (accessed November 13, 2023).

Williamson, K., A. Satre-Meloy, K. Velasco, and K. Green. 2018. *Climate Change Needs Behavior Change: Making the Case for Behavioral Solutions to Reduce Global Warming*. Arlington, VA: Rare.

Wood, W., and D. T. Neal. 2016. "Healthy through Habit: Interventions for Initiating and Maintaining Health Behavior Change." *Behavioral Science and Policy* 2 (1): 71–83.

6

Concluding Remarks: Looking Back and Going Forward with EPR

David Nightingale
Special Waste Associates, Olympia, Washington

To understand where we might want, or need, to consider going with Extended Producer Responsibility, it is useful to briefly take stock of some of the materials management practices of our recent past.

Prelude to EPR

As a boy, I vividly recall being constantly on the outlook for the errant tossed, but intact, glass beverage bottle so that I could bring them into my local grocery store and claim the 5 or 10 cent per bottle deposit. The popular term "recycling" was new then. In my parents' era, there was an aggressive nationally coordinated system to recover copper, steel, aluminum, and other commodities for the Second World War materials supply system. And earlier, during the Great Depression, nobody threw anything of value away and everything possible was recycled or reused to the maximum extent. These materials management systems were mandated by the federal or state government and driven by the economic self-interest of industry and individuals.

An exotic innovation of the time was the fluorescent light bulb. Although expensive, they were an amazing new-science technology that saved massive energy compared to incandescent lamps. Retail establishments that switched to fluorescent lamps were seen as modern and progressive pioneers. I was impressed when my grade school in Los Angeles did the same. There was no discussion about what happened at the product end-of-life for these fragile, mercury-laden, glass tubes. Only later would the toxicity of mercury in fluorescent bulbs become known. Historically, most machines, appliances, fixtures, paper, and other commodities had components that could be recycled, composted, or were readily degraded in a landfill.

These times also reflected when there was little in the way of synthetic chemicals in the home. For antiseptic cleanliness, a weak bleach or ammonia solution was the cure. As synthetic chemicals (such as halogenated hydrocarbons and new pesticides) were increasingly introduced into the mass retail market, solid waste operators, wastewater treatment operators, and environmental organizations began seeing these persistent and non-degradable chemicals in landfill leachate, waste-to-energy emissions, as well as wastewater effluent and sludge. In retrospect this should have been anticipated, but at the policy and regulatory levels this awareness did not start emerging until the mid-century or later.

The initial response by regulators was to go after the largest contributors of these and other difficult-to-manage chemicals in waste streams: the commercial and industrial sources. Control of household sources for these chemicals was not part of the regulatory scheme. Not until the mid-1980s was there an emerging awareness that households and small businesses were also significant contributors to this problem. Hence, the development of local household hazardous waste (HHW) collection programs for materials such as cleaning liquids, pesticides, and other chemical products with hazardous properties. But HHW collection is expensive and local governments constantly struggle to find funding for all of their important programs, let alone discretionary programs, which HHW collection is in most states. This was one of the primary initial drivers for EPR in the United States: the gap between a need to manage these problematic chemicals and restricted local and state government budgets to do so.

Next Step: The EPR Stage Is Set

As with most complex problems, there are multiple available paths to a solution. Looking back at historic materials management systems, as well as what has been done more recently in what we now call Extended Producer Responsibility (EPR), helps us think about a different future. In the chapter on EPR by Sanborn, Goncharoff, and Wells, some sage advice is provided on what has worked and been problematic in passing and implementing EPR laws during the past fifteen years or so.

The Product Stewardship Institute, the CA Product Stewardship Council, and the NW Product Stewardship Council Steering Committee were early leaders in promoting EPR across the United States. These EPR pioneers "broke the ice" with progressive states and municipal governments passing local, statewide, and regional EPR policy and laws. Due to these EPR leaders, industry allies, and others, programs for consumer rechargeable batteries, leftover architectural paints, and electronics were launched and grown across the United States. No longer was EPR the "new, untested, and experimental vehicle."

Materials management systems have often been mandated by state and federal law. In general, commercial interests understand that governments have historically imposed conditions in the marketplace for the public good and have a legitimate stake and authority to act in this arena. The Pharma case that went all the way to the Supreme Court (mentioned in the chapter by Sanborn and Goncharoff) sent a national signal of the legitimacy of the EPR approach. This was in essence, a re-acknowledgment of the well-established government obligation to protect the health and well-being of its citizens.

The history of government-mandated materials management systems is often needed where the intrinsic value of waste materials does not drive reuse, recovery, and recycling in combination with a recognized public health risk or important public benefit. HHW generally falls into this category of negative market value materials and a recognized public health risk. HHW also plays in the public benefit category regarding climate change due to the need for recyclables that are less contaminated.[1] In 2023, Vermont became the first state

[1] Luminary Interview: Debbie Raphael—Connection between HHW and Climate Change, *Waste Advantage Magazine*, October 2021. https://wasteadvantagemag.com/issues/#flipbook-october-2021/7/ (accessed 2023).

to enact EPR legislation (H.67) for statewide comprehensive HHW collection and disposal. Previously, only certain subsets of HHW were covered by any state EPR law.

In the past few years EPR laws have grown in application to include packaging, plastics, solar panels, and other materials that, at their end of life, are problematic to manage. Given those facts, what might be the possible EPR future for other wastes—resources—in the United States?

Nationwide EPR in the United States

In the early 2000s, during the Paint Product Stewardship Initiative led by the Product Stewardship Institute, the Paint and Coatings Association representatives indicated that they would not pursue a national EPR law for leftover architectural paints until there was a critical mass of states with such laws in place. With twelve states and the District of Columbia with paint product stewardship laws on the books, has a required critical mass been attained?[2]

The answer appears to be, "not quite yet." While great progress has been made on EPR for architectural paint, more states need to come on board. Looking at the locations where paint EPR laws exist, it isn't an even geographic distribution. In the eastern time zone, no state south of Maryland has a paint EPR law. In the central time zone, only Minnesota and Illinois have paint EPR in the books. In the mountain time zone, only Colorado has legislated a paint EPR law. West of the Rockies, only California, Oregon, and Washington have paint EPR programs. As of the 2023 census estimates, these jurisdictions represent about 32.4 percent of the US population.[3] Without paint EPR laws in a number of southern, midwestern, and Rocky Mountain states, it would be surprising to see a national paint EPR law passing in the US Congress.

[2] The jurisdictions with paint EPR programs are Rhode Island, Oregon, Washington, California, Colorado, New York, Connecticut, Minnesota, Vermont, Maine, Illinois, Maryland, and the District of Columbia. https://productstewardship.us/products/paint.

[3] Based on official US Census Bureau 2023 population estimates, the states with PaintCare laws are California, Colorado, Connecticut, Illinois, Maryland, Maine, Minnesota, New York, Oregon, Rhode Island, Vermont, and Washington, plus the District of Columbia.

On the other hand, electronics (e-waste) EPR laws have been adopted more broadly with twenty-four states and the District of Columbia in that list. The eastern seaboard is well represented from Maine all the way to South Carolina, except for New Hampshire, Massachusetts, and Delaware. West Virginia and Pennsylvania also have e-waste EPR laws. Similarly, the southern Plains and Midwest have a significant number of states with e-waste EPR including Texas, Oklahoma, and Missouri and five states around the Great Lakes. In the west, Washington, Oregon, California, and Hawaii have all passed e-waste EPR legislation.[4] This means that 67 percent of the US population resides in a state with existing e-waste EPR programs. The biggest regional gaps are in the Rocky Mountains, Great Plains, and Deep South. Florida, Georgia, Arizona, and Tennessee are the four largest population states without e-waste EPR.

It might appear that with just over two-thirds of the population in e-waste EPR states, it could be a good candidate for a national e-waste bill in Congress. However, with just under half of the states represented, it might be a tough sell in the Senate, where population differences are mitigated. From an implementation standpoint, the existing e-waste EPR laws vary considerably between the states. This lack of uniformity of existing programs makes it challenging to formulate a national law. A national law would have to be negotiated to override or accommodate the existing variations. This complicates the process. So, it's not clear if there is yet a critical political mass to pass a national e-waste EPR law. Still, we are heading in the right direction with steady growth in EPR legislation across the United States (see Figure 6.1). In the not-too-distant future, there may be an impressive catalog of states that make a national uniform law desirable for several waste categories. A push for national EPR could be enhanced by national and regional manufacturers and producers that would much prefer a single set of regulations and performance requirements, rather than a patchwork of state-by-state laws.

In addition to EPR for paint and e-waste, national EPR laws addressing used motor oil, pharmaceuticals, mercury-containing devices, gas cylinders, and other HHW categories are some of the possible candidates for nationwide

[4] A map showing the states with e-waste EPR is found here: https://productstewardship.us/products/electronics.

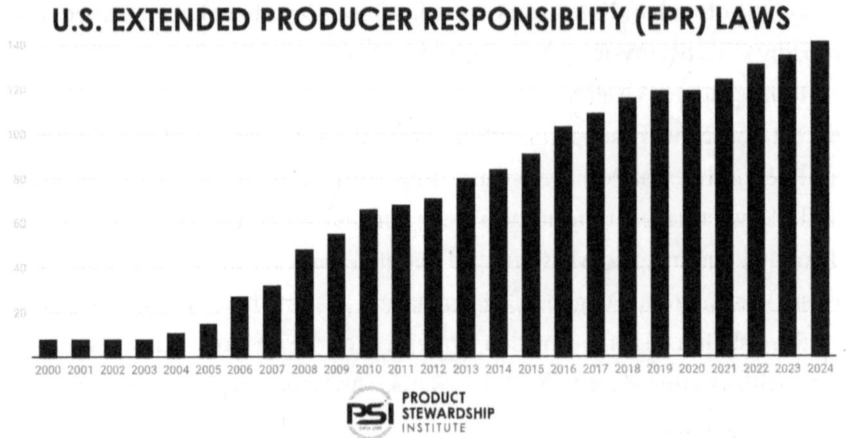

Figure 6.1 US Extended Producer Responsibility (EPR) laws as of 2024.

Source: Product Stewardship Institute, 2025.

EPR legislation. Recently, beyond the traditional HHW categories, there has been mounting interest in national EPR laws for packaging and disposable wipes. This may increase the receptiveness of EPR as a model for all types of HHW and other problematic materials in the waste stream.

The relationship to the materials recycling system and climate change might move Congress to consider national EPR laws, including HHW, as discussed in the following paragraphs.

China's National Sword Policy, HHW, and Climate Change

In the past few years there has been a scramble to find or develop markets for many traditional recycled waste streams that were being exported to China before China's National Sword policy. Implemented in 2018, the National Sword policy severely curtailed the amount of waste and low-quality recyclables it imported for the United States and Europe. This blunted the development of the US recycling system that increasingly generated large volumes of lesser quality, greatly contaminated recyclable materials.

About twenty years ago, most recycling systems were not "single stream" (commingled materials). Rather they consisted of separated materials such as white office paper, glass, metals, newsprint, corrugated cardboard, and plastics. These were separated into multiple containers for curbside and recycling drop-off centers before preparation for final markets.

Single-stream recycling relies on recycling collection bins, sorting lines, and automated processing to increase the collection and processing cost-effectiveness. This increases the volume but typically decreases the quality of the recyclables in the end. When the Chinese market effectively disappeared in 2018, due to low-quality, contaminated recyclables, the US recycling market problem became acute.

The National Sword policy imposed a "market correction." in Wall Street parlance. To meet the new market conditions, recycling streams must literally be cleaned up with excess contamination of the recyclables avoided, to find willing takers of materials. One of the contaminants found in materials recovery operations is HHW. To the extent that HHW is effectively removed from the municipal waste stream, recyclable materials quality improves and increases marketability of the recycled commodity.

A direct benefit of being able to recover and find markets for more recyclable materials is a systematic reduction in energy use and carbon emissions. In that way, recycling and HHW collection programs contribute to a more circular economy, a common goal of EPR, and help reduce emissions of climate change gases.[5] Climate change is important to many lawmakers and members of Congress from states without EPR legislation.

The Future of EPR Organizational Structures

The design, scope, and flexibility of EPR programs is not a fixed formula. To operate programs that are most cost-effective, economies of scale serve as a significant driver. When you can use the same infrastructure (labor, collection, transportation, and storage) to manage multiple types of materials, your fixed

[5] This connection between proper HHW management, the circular economy, and climate change was described in an interview with Debbi Raphael. https://wasteadvantagemag.com/issues/#flipbook-october-2021/7/.

and variable operating costs may be significantly reduced. Applying these economic principles to HHW EPR program design leads to questions about how to optimize future EPR systems.

Development of national EPR laws can provide producers and manufacturers with the opportunity to reach national economies-of-scale. There is the potentially compounding economic advantage of coordinating EPR programs for multiple types of HHW or HHW-related materials. An abbreviated case study of what was originally the British Columbia (BC) Paint Care Association (now the Product Care Association), provides a complex and illustrative example of adaptation, expansion, and combination of various EPR programs over time.

In 1994, BC Paint Care Association was incorporated as a nonprofit organization to manage leftover architectural paints in British Columbia, Canada, within an EPR legal construct. The BC Provincial government subsequently added HHW in 1997 for which the obligated industry formed the nonprofit organization, Consumer Product Care Association. In 2001 BC Paint Care Association and Consumer Product Care amalgamated to create Product Care Association.

Product Care Association then began expanding across Canada, adding provinces and programs for paint, lighting, alarms, and HHW and providing management services to other EPR programs from 2001 through today. Programs now span across Canada including BC, Saskatchewan, Manitoba, Ontario, Quebec, New Brunswick, Nova Scotia, Newfoundland, and Prince Edward Island, with additional programs in Alberta and Yukon Territory in the planning stages.

In addition, a 501 (c)(3) PCA-USA was formed to provide programs in the United States. This has included working with US PaintCare on the paint EPR program in Oregon (2009–14), LightRecycle Washington (state) (2015–), California Thermostats (2022–), and work with the US Mattress Recycling Council.[6]

There have been discussions and hints of combined legal structures or frameworks in the United States, but in general this coalescence of EPR types

[6] Personal communication with Mark Kurschner, Product Care, January 23, 2023.

and coordinated administration and operations have been thus far lacking in the United States. The exception to this is the work being done by PCA-USA and the nationally coordinated PaintCare system mentioned in Chapter 2 by Jim Quinn.

An early attempt to create a coordinated legal structure in the United States between different product groups is found in legislation from Washington state in 2006.[7] This law mainly focuses on management of e-waste but created the more broadly named Materials Management and Financing Authority. This name was intentionally used to provide an avenue to combine multiple EPR types under one administrative structure. However, subsequent EPR bills in Washington state have yet not leveraged this opening.

The evolution from the BC Paint Care Association to the Product Care Association may portend a more efficient and coordinated US EPR system of the future. Certainly, the evolution of EPR and how it will ultimately be administered and implemented is a story yet to be told. It is anticipated that ways can be found to coordinate compatible pieces of the EPR landscape to provide a more efficient and effective system of EPR. It is this direction in which we are heading, to the benefit of many and varied stakeholders.

[7] Washington materials management and financing authority at https://app.leg.wa.gov/RCW/default.aspx?cite=70A.500.270.

Appendix: MetroPaint and Green Building Fact Sheet, 2014

MetroPaint and Green Building

Green building focuses on a variety of ways to make the construction and operation of buildings more environmentally sound and sustainable. MetroPaint is an excellent choice for many green building projects, whether they are new construction, renovation or maintenance on housing, or commercial and industrial buildings.

This information sheet addresses MetroPaint's performance on several environmental fronts:

- Volatile Organic Compounds (VOCs)
- LEED™ and other green building certification systems
- Life cycle benefits

Green building, per Wikipedia, is "The practice of increasing the efficiency with which buildings and their sites use and harvest energy, water and materials, and reducing building impacts on human health and the environment, through better siting, design, construction, operation, maintenance, and removal—the complete building life cycle."

Volatile Organic Compounds

Many green builders are concerned with the VOC content of the paint they use, as they are an important consideration for both environmental and human health impacts.

What are VOCs? VOCs are a class of carbon-based chemicals that have the capacity to evaporate readily at room temperature. They can degrade air quality both in buildings and in the atmosphere. They may produce odors, and they contribute to smog. The health effects of different VOCs vary considerably. The four VOC compounds that are used in the production of latex paint (ethylene glycol, propylene glycol, texanol and butyl carbitol) are all considered to have low toxicity, though even these can cause nose and throat irritation and other discomfort at high concentrations.

What are EPA limits on VOCs in paints? The US Environmental Protection Agency (EPA) air quality regulations (40CFR59.400 et. seq.) limit the VOC content of paint. These limits are intended to protect the environment by reducing the creation of photochemical smog in the atmosphere. For the class of paints that includes MetroPaint, the VOC content is limited to 250 grams per liter.

What is the VOC content of MetroPaint? In 2013, Metro tested batches of every color of our paint for VOCs. The VOC content averaged 91 grams per liter (g/L), and no batch was over 118 grams per liter, well under the EPA limit.

What are "Low-VOC" paints? Some manufacturers have developed latex paint products that are lower in VOCs. However, there is no regulatory definition of "Low-VOC," so the VOC content of paints labeled as Low VOC can vary. Some are simply in compliance with EPA regulations. Others may be significantly lower, some are even close to zero and may be labeled as "No VOC" or "Zero VOC." Green Seal™, an international standards-setting organization for environmentally preferable products, has established VOC limits for virgin paints in standard GS-11 (www.greenseal.org). Under this standard, flat paints with colorants added must have less than 100 g/L VOCs.

When are Low-VOC paints a good choice? When a building interior is to be painted and there are concerns about indoor air quality, particularly among individuals who are sensitive to odors or chemicals during the first few days after painting. The most significant concentration of VOC vapors occurs within 100 hours after painting.[1]

[1] An EPA study concludes that the rate of off-gassing of the four VOC compounds found in latex paint can vary considerably depending on the substrate. For non-porous surfaces such as stainless steel, nearly all of the VOCs are emitted within the first 100 hours after painting. In contrast, freshly painted porous surfaces, such as wood or sheetrock, release a relatively small amount of the VOCs in the initial "puff." They emit the remaining VOCs at very low levels over time, possibly as long as 3.5 years. However, the study found that even on porous surfaces, the ethylene glycol vapors dropped to under 1 mg/m^3 by 100 hours, and down to about 0.1 mg/m^3 for most of the long off-gassing process. For comparison, the Occupational Safety and Health Administration allows occupational exposures of ethylene glycol up to 100 mg/m^3. The other three VOCs were found in even smaller amounts after 100 hours, close to the lower limit of the test method's ability to detect (Chang, 2001).

When is MetroPaint a good choice for the environment? MetroPaint's modest VOC content is appropriate for exterior applications and interior applications other than those with indoor air quality concerns. But VOCs are not the only environmental consideration—there are a number of other environmental benefits provided by recycled paint (see the section "Life cycle Benefits").

Additional Notes about VOCs

- The VOC levels in EPA regulations and Green Seal's virgin paint standards are intended to apply to paint before it is tinted at the point of sale. Colorants added at the point of sale contain VOCs as carriers. Darker colors in particular can have a lot of colorant added, and thus add a lot of VOCs. Recycled paint is at a disadvantage here as post-consumer paint contains these colorants. Tinting can add as much as 80 g/L to the VOC content. So the VOC content of MetroPaint would be somewhat lower if it could be calculated the same way.
- Under EPA's air quality regulations, latex paint recycling is encouraged, and recyclers are allowed to use an "adjusted VOC content" calculation, taking into account the percentage of paint that is recycled from post-consumer sources. If MetroPaint used this calculation, our adjusted VOC content would be between 0 and 68 g/L.
- The Green Seal standard for recycled paints, GS-43, includes the requirement that the VOC content of recycled paints be under 250 g/L. MetroPaint complies with this limit.
- EPA's VOC regulations are targeted at atmospheric pollution, and not all harmful chemicals are included in their definition of VOCs. So a paint that is labeled as Low or No VOC may still emit some vapors that are harmful to human health. A 2001 EPA study found that "certain paints marketed as Low VOC may still emit significant quantities of air pollutants" (see footnote 1). In some cases, the overall quantity of VOCs was quite low, but significant amounts of specific compounds were nevertheless emitted.
- The VOCs in recycled paint have already been manufactured and will enter the environment one way or another. When recycled paint is used instead of virgin paint, recycled paint reduces the need to manufacture new VOCs.

LEED and Other Certifications

LEED, or Leadership in Energy and Environmental Design, is a benchmarking system developed by the US Green Building Council (www.usgbc.org) to certify high-performance green buildings. Building projects accumulate points toward certification in a variety of different categories. Use of MetroPaint can help garner points for LEED certification in a couple of ways:

- Under the Recycled Content credits (MR4.1/4.2), use of recycled paint can help qualify for credits.
- Under the Regional Materials credits (MR5.1/5.2), use of paint that is manufactured regionally helps gain credits. This is defined as within 500 miles, which includes much of the Pacific Northwest for MetroPaint's Portland-based plant. In LEED for Homes, credit is specifically awarded for use of recycled paint that is certified by Green Seal. MetroPaint is the first paint recycler in the world to offer Green Seal certified recycled paint. And in the new LEED v. 4, under BPDO (Building Product Disclosure and Optimization), Sourcing Raw Materials Option 2, credit is given for materials manufactured by companies that show Extended Producer Responsibility. That includes MetroPaint.
- There are other green building certification systems, such as the guidelines developed by the National Association of Home Builders® (NAHB) for residential construction (www.nahb.org), and the Green Globes system (www.greenglobes.com). These guidelines are similar to LEED in that points are available for both recycled content building materials and regionally- manufactured materials.

Life Cycle Benefits

For years environmental advocates have been promoting the recycling of various household commodities, such as newspaper, glass, aluminum, etc. Recycling these materials helps reduce the amount of water, raw materials, and energy used. Less pollution is generated and less landfill space is used. All of these benefits hold true for recycled paint as well.

Production of new paint generates pollution during the extraction of raw materials, the manufacturing process, and the transportation of final products. Every time a gallon of MetroPaint is used instead of a gallon of new paint, a number of benefits are obtained:

- **Significant energy savings and greenhouse gases are reduced.** Embodied energy is the amount of energy it takes to produce a good or service. The embodied energy per pound in new paint is greater than that of any other common construction material, including steel, glass, or lumber. This means that when recycled paint is used instead of new paint, significant amounts of energy are conserved and significant reductions in greenhouse gases are achieved. Depending on the assumptions used, life cycle studies of paint show that use of recycled paint can achieve an energy reduction of up to 90 percent, eliminating up to 115 pounds of carbon dioxide emissions for every gallon of recycled paint used.
- **Various types of pollution are not generated.** A significant percentage of the ingredients in latex paint is derived from petrochemicals. The extraction of these chemicals results in various environmental impacts, notably air emissions and waste discharges to surface waters. The extraction of titanium dioxide, another important latex paint ingredient, results in the production of significant quantities of sulfuric acid, metal sulfates, and metal chlorides. These wastes are commonly disposed of in the ocean or via deep well injection. A 2009 life cycle study, conducted jointly by the paint industry and government agencies, confirmed that recycling of paint provides significant reductions in a variety of impacts to human health and the environment, such as carcinogens generated, respiratory effects, ecotoxicity, eutrophication, and acidification.
- **Less water and fewer raw materials are used.** The extraction of ingredients, the manufacturing process, and transportation of virgin paint require substantial raw materials and water. The relatively simple recycling process uses significantly less of these. The government or industry life cycle assessment confirms that significant reductions in water usage and mineral extraction are obtained with paint recycling.
- **Landfill space is conserved.** At MetroPaint Swan Island, more than 600,000 containers of paint are opened each year. Paint that is recyclable is made into our recycled product, and steel and plastic paint cans and pails

are sent to recyclers. This keeps more than 1,000 tons of paint and more than 100 tons of containers out of landfills each year.

The Bottom Line

MetroPaint provides a number of environmental benefits and can help green building projects get certified. The environmental benefits are so substantial that the independent environmental certifying organization Green Seal has certified MetroPaint.

Reference

Chang, John C. S. 2001. "Capstone Report on the Development of a Standard Test Method for VOC Emissions from Interior Latex and Alkyd Paints." US EPA 2001. EPA/600/R-01/093.

Note

Reprinted with permission: The wording herein is from an existing fact sheet. https://www.oregonmetro.gov/sites/default/files/2014/06/02/metropaint_and_green_building_fact_sheet_2014.pdf.

Contributors

Kristin Aldred is an environmental psychologist and systems-based sustainable business practitioner. Throughout her career she has used knowledge of people, design, and strategy to understand what influences human behavior and then pursue changes in the design of products, environments, processes, and policies for human well-being and the health of our environment. She has held roles in environmental conservation, corporate responsibility, policy, business strategy, and research in the private and nonprofit sectors. Aldred completed a doctorate in Human Behavior and Design from Cornell University and holds a master's degree from Oregon State University's College of Forestry and a bachelor's degree from Wesleyan University.

Tim Goncharoff is managing principal for Clements Environmental and the National Stewardship Action Council board president. He is a passionate, creative, and dedicated advocate for the environment and public health, writing and speaking frequently about Zero Waste, Extended Producer Responsibility, and the circular economy. Goncharoff has extensive experience in the public, private, and nonprofit sectors and is well-known for developing and implementing cutting-edge EPR ordinances and programs for years a leader of the groundbreaking waste reduction and environmental protection efforts of Santa Cruz County, California. His milestones include passage of the first and only EPR ordinance for medications and sharps in the United States, the first ordinance to eliminate the use of single-use plastic toiletries packaging in the hospitality industry, and the first in the nation ordinance to eliminate single-use plastics in fast-food packaging. Goncharoff is the recipient of many awards, including the Governor's Environmental and Economic Leadership Award, California's most prestigious environmental honor.

The **Mattress Recycling Council (MRC)** is a nonprofit organization formed by the mattress industry to operate recycling programs in states that have enacted mattress recycling laws. The program is currently running in California,

Connecticut, and Rhode Island, and its expansion to Oregon is expected. MRC is committed to educating consumers and the mattress industry about state mattress recycling laws and assisting retailers and manufacturers with compliance. Through its work with local governments, waste management professionals, recyclers, and others, MRC has successfully developed accessible and efficient collection and mattress recycling networks. Contributors to Chapter 4: Mike O'Donnell, chief operating officer; Justine Fallon, vice president of operations; Kate Caddy, sustainability director; Marty Salamone, coordinator of Policy and Government Affairs; Marie Clarke, formerly vice president of Industry and External Affairs.

Tom Metzner has been an environmental analyst with the Connecticut Department of Energy and Environmental Protection for over thirty years. He works on solid and hazardous waste policy including Connecticut's existing Extended Producer Responsibility programs and Toxics in Packaging law. He has served on the board of directors for the Product Stewardship Institute including as president. He chaired a legislative working group on microfibers, and he represents Connecticut in the Toxics in Packaging Clearinghouse and the Interstate Chemicals Clearinghouse.

David Nightingale began his household hazardous waste career as a consultant in the late 1980s, when he wrote local hazardous waste plans for counties in Washington State. He now owns the consulting firm Special Waste Associates in Olympia, Washington, which assists local and state governments in planning and development of household hazardous waste (HHW) management systems. Previously, while working for the WA Department of Ecology, Nightingale represented the NW Product Stewardship Council (NWPSC) and Washington state for the five years of the National Paint Product Stewardship Initiative. This led to the formation of PaintCare and a series of state paint EPR laws. Subsequently, he led the NW Product Stewardship Council's Electronics initiative. Nightingale worked with the publisher of *Waste Advantage Magazine* to compile articles from his "HHW Corner" column into the book, *Chronicle of the HHW Corner*, which includes articles on Extended Producer Responsibility. Nightingale is a board member of the North American Hazardous Management Association.

Jim Quinn is a resilience planner for Metro, the regional government in the Portland, Oregon area, where he focuses on the resilience of the region's solid waste system. Prior to that Quinn worked in the hazardous waste management field for thirty-four years, including twenty-three years managing the household hazardous waste program and paint recycling program at Metro. He holds a bachelor's degree in chemistry from Reed College, and a master's degree in environmental management from University of San Francisco. Quinn is also a board member and Policy Committee chair for NAHMMA, the North American Hazardous Materials Management Association.

Heidi Sanborn serves as the founding executive director of the National Stewardship Action Council, a 501(c)4 that advocates for a circular and equitable economy through Extended Producer Responsibility policies. She has been in the waste management industry for more than thirty-two years working as a consultant, regulator at CalRecycle serving as policy advisor to the chair in 2000–2, and advocate for producer responsibility and a circular economy as the founding director of the California Product Stewardship Council, a 501(c)3 and the National Stewardship Action Council. In May 2020, Sanborn was appointed by the California Environmental Protection Agency to the Statewide Commission on Recycling Markets and Curbside Recycling and elected chair. She has led coalitions and sponsored bills that are first in the nation EPR and Truth in Labeling bills in California and was a key voice in helping Senator Ben Allen in 2022 to pass the first true Circular Economy bill in the world that includes robust EPR for packaging combined with environmental mitigation funding.

Jennifer Tabanico is the president and owner of Action Research, a firm that specializes in changing behavior for the public good by applying marketing and social science research to outreach programs that promote safe, healthy, and sustainable communities. Tabanico currently oversees behavior change programs addressing water quality protection, human–wildlife interaction, food waste reduction, and recycling. She also serves as an editorial board member for *Social Marketing Quarterly* and as an instructor for the Behavior Change and Sustainability Certificate Program for University of California San Diego (UCSD) Extended Studies. Tabanico has a master's degree in experimental psychology from California State University, San Marcos.

Index

Note: Figures are indicated by page number followed by "f". Tables are indicated by page number followed by "t"